Einstein, et al.

**Manifestation, Conflict REVOLUTION® &
The New Operating System**

Barbara With

Einstein, et al.
Manifestation, Conflict REVOLUTION® & The New Operating System

© 2016 Barbara With
ISBN: 978-0-9910109-3-6
Library of Congress Control Number: 2016903012

All rights reserved. No part of this book may be used or reproduced
in any manner whatsoever without written permission.

For information, address
Mad Island Communications LLC
P.O. Box 153
La Pointe, WI 54850
715.209.5471
barbarawith11@aol.com

www.barbarawith.com
www.partyof12.com
www.barbwith.com

Book Design: Barbara With
Cover photo: Oren Jack Turner, Princeton, N.J., Library of Congress, Public Domain
Editors: Sandy Grabowski, Debbie DeLung, Mary Catherine Kline, Robin Cordova
Photo of Barbara on Map and author photo: Brenda Miles, www.brendamilesphotography.com/

Acknowledgments

A passionate thank you goes to the participants of the group channels, workshops, classes, readings, focus groups, and discussions who have been a part of the research and development of Conflict REVOLUTION® and purveyors of these Einsteinian thoughts.

My unending gratitude goes to those who have stood by me on my journey as Einstein's emissary and have taken his instruction as your own. Without you, I would be lost. Without your willingness to participate in the grand experiment and imagine that it really is Einstein, I would be walking alone on this magnificent and often times mind-boggling path. You have taught me how to not get stuck in the same patterns of defending myself from the opinions of others and blaming them for my problems. Because of our work, I understand how to deeply feel Emotion, voluntarily revolve my conflict, and take action for the good of all. All of you are a part of a rare and miraculous experiment that may not be fully understood until many years after our passing.

And to Einstein himself, he who endures years after his death as a living, beloved symbol of what is possible. His dedication to peace is still a shining voice in a dark time; his insatiable and brilliant hunger for truth continues to drive an era forward. Not a day goes by when I am not in awe of this mission, and that I volunteered and was chosen to scribe this noble ideal.

Dedication

This book is dedicated to my mother.

Table of Contents

Introduction — Einstein, Conflict REVOLUTION® and Today i
Barbara With

Prelude — The New Operating System 1
The Party

Part I — The Science of Compassion 7
    The Origins of the Vision 9
    The Human Intention 25
    The Three Human Dimensions 33
        Emotion 39
        Intuition 45
        Intellect 49
    Honored 4th 61
    Abscesses 65
    The Human Body 69

Part II — Conflict REVOLUTION® Learning Project 77

Part III — The Action Plan 121

Finale — Living Aligned to Compassion: Cn^3 139

About the Author — Barbara With 149

Barbara With & Flat Einstein
Collège de France, Paris
December 2014
Photo: Barbara With

Einstein in Paris

In 1922, Albert Einstein traveled to Paris to deliver a talk on his newly formulated Theory of Relativity to an anxious group of listeners at the Collège de France. In his previous appearances in the US, England, and Italy, Einstein chose to speak in German and Italian, respectively. When he arrived in France, according to physicist and astronomer Charles Nordmann, he was nervous about his lack of fluency in French. He decided to do a short dissertation before opening for questions and conversation. This would allow him to navigate through the language more thoughtfully and accurately.

"The theory of Einstein is a marvelous tree that has grown farther and higher than any other ideal flowers of human thought," Nordmann concludes.

Einstein's Paris mission was more than just to inform the French of his revolutionary scientific ideas. On June 28, 1919, Germany and the Allied Nations—Britain, France, Italy, and Russia—signed the Treaty of Versailles, formally ending World War I.

Oddly, Einstein biographies often herald his trip to Paris as one the most important events of the era—not because of the science, but as a peace mission that began mending ties between Germany and France and reestablishing international intellectual relations that had been cut off during the war.

Einstein's presentation was attended mostly by scientists, researchers, and philosophers. There is no documentation of his speeches from this presentation; however, Nordmann took notes on the event and published them as *Einstein and the Universe* later in 1922.

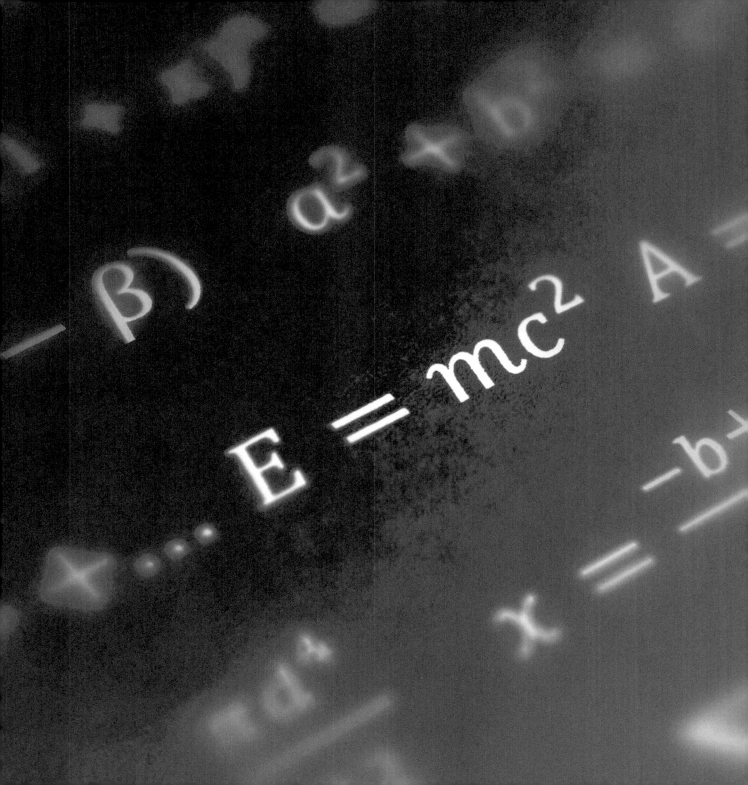

Introduction

Einstein, Conflict REVOLUTION® and Today

Channeling Einstein, et al.
Quantum Leap, Taos, NM
September 19, 2007
Photo: Dave Schemel

Introduction

Barbara With

Since 1987, I have been conducting and documenting hundreds of channeling sessions around the world, speaking what I believe are the voices of Albert Einstein and other inventive, dedicated peacemakers from beyond the grave whom I refer to as the *Party of Twelve* or the Party. They say they are here to teach a process of "world peace, one person at a time, starting with the individual self," and they need our participation in the experiment to test their theories.

This volume is not about how I came to be Einstein's emissary; that is documented in *Party of Twelve: The Afterlife Interviews*. Suffice it to say, it's been a wild ride. More importantly, I want to share what I have learned from studying my Imagined Einstein's theory known as Conflict REVOLUTION® or Con Rev. Having heard his brilliant worldview in thousands of sessions over almost 30 years, I successfully taught myself to make manifest his foolproof process to resolve conflicts from within. The outcome has been exactly as he predicted in the beginning: resolve conflict on the root level of the human psyche and the resulting manifestation naturally arises for the good of all. When peace is achieved within, it cannot help but manifest in the physical world.

The results have been what Einstein calls *installing a new operating system*, wherein we are intuitively guided in any given moment to take control of our own Domain, where our power truly lies. Using what Einstein calls the three *Human Dimensions*—Emotion, Intuition, and Intellect—we can learn how to use free will to get out of the way and allow Compassion, the fifth fundamental force of the universe, to naturally manifest in our lives for the greatest good of the entire situation, whatever it may be.

When we have the courage to use this regenerative process, we can manifest a resolution for the greatest good without applying pressure directly in the arena of the conflict. Einstein did say, after all, that we cannot solve our problems at the same level of thinking at which we create them. Once we quit fighting in the physical world and instead turn inward and resolve conflicts at their root—between the three Human Dimensions—we literally influence how matter manifests.

For much of my life, I was a highly creative but very angry woman. Constantly victimized, ensnarled in dramas, manifesting conflicts at every turn, I was literally asking for it on a daily basis. I picked fights, suffered intense mood swings and chemical addictions, and eventually was challenged with mental health issues that no amount of counseling seemed to transform. Slowly, following Einstein's step-by-step instructions, I learned empirically where my power is and eventually changed my knee-jerk reactions. Debilitating conflict became creative, inspired solutions that then had a profound and positive impact on outcomes. By installing this new operating system, I greatly reduced the need to haggle over issues with people; instead I can get to the root of my own pain and lovingly process it before dragging anyone else into a drama. This is peacemaking.

Yes, I'm still human and still struggling with conflict, but much of the backlog of pain I carried throughout my life has dissipated. The new operating system gives me a way to keep Emotion moving through me as I take control of my thinking, making intentional, conscious decisions based on intuitive inspiration about what is being created in any given moment. This is what Einstein calls being *aligned to Compassion* (consciousness to the third power, or Cn^3) and the powerful position of manifestation that he describes in our first book, *Imagining Einstein: Essays on M-Theory, World Peace & The Science of Compassion*.

After that book was released, the search began for someone to create the maps of human consciousness that he has described ad infinitum. I asked painters, sculptors, graphic artists, programmers, anyone who might come up with a pictorial or even 3D articulation of his maps.

Introduction

In the end, I hunkered down one winter and created them myself using my own limited design skills, mentored by my Imagined Einstein.

Einstein, et al. is a compilation of those maps, transcripts from the many sessions we have conducted since 1993, excerpts from *Imagining Einstein,* and empirical experience that I and others have acquired putting this new operating system into practice in our own lives. The voice of "we"—Einstein, the Party, you, me, et al.—represents all who are working together towards the evolution of the entire human species.

This book is about how to take control of your Domain, install this new operating system, and inspire yourself to use it to make your own miracles. In the process, you redefine the term "miracle" and understand that gaining power over yourself is not only your right, but your responsibility—an obligation you have to yourself, your family, the communities in which you live, and to the continued survival of the planet.

During the 1920s, Einstein made many presentations throughout Europe, not just about his revolutionary scientific theories, but inspiring the world to come to peace after World War I. Many of the photographs included in this book are from my trips to Europe, when I followed in his footsteps through Paris, Madrid, Prague, and Copenhagen. I visited the places he had appeared during those years, as well as local cemeteries. The tombstone of Yolanda Gigliotti on page 2 was a lovely surprise. I had no idea who she was when I snapped the picture at Montmatre Cemetery in Paris, other than an artistic soul with a striking tombstone.

Born January 17, 1933, Ms. Gigliotti was also known as Dalida, "an Egyptian-born, Italian-French singer and actress who ranks among the six most popular singers in the world, receiving 70 gold records throughout her career."[1] On May 2, 1987, she took her own life, leaving a note behind: "La vie m'est insupportable ... Pardonnez-moi." ("Life has become unbearable for me ... Forgive me.")

---

1 https://en.wikipedia.org/wiki/Dalida

What is even more ironic is the tombstone pictured on page 147. As the book was nearing completion, I needed a few more photographs, one in particular to be part of a meme on creativity. I chose a unique tombstone with an unusual cast of the face of the deceased, as he smoked his pipe well into eternity. As it turns out, Guy Pitchal had been Yolanda's psychotherapist!

Let Yolanda, Guy, and the honorees of all the other headstones strewn throughout these pages serve as a gentle reminder that life is fleeting and Afterlife awaits us all.

While preparing this book for press in February 2016, a phenomenal discovery came to light. Scientists at the Laser Interferometer Gravitational-Wave Observatory (LIGO), run by the California Institute of Technology and the Massachusetts Institute of Technology, announced that they had detected the gravitational waves first predicted by Einstein in his Theory of General Relativity in 1915. LIGO studied two black holes colliding and documented the gravitational waves with sensitive monitoring equipment able to detect fluctuations in the spacetime continuum. A press release by Caltech and LIGO announced:

> For the first time, scientists have observed ripples in the fabric of spacetime called gravitational waves, arriving at the earth from a cataclysmic event in the distant universe. Physicists have concluded that the detected gravitational waves were produced during the final fraction of a second of the merger of two black holes to produce a single, more massive spinning black hole.[2]

I almost passed out when I heard this news. You mean, science has detected *our* gravitational waves? The waves that my Imagined Einstein, et al. have been working with for over 20 years? We have intricate, intimate maps not only of the waves, but of human consciousness and the

---

2   "Gravitational Waves Detected 100 Years After Einstein's Prediction," LIGO Hanford Press Release, February 11, 2016 https://www.ligo.caltech.edu/news/ligo20160211

Introduction

entire operating system, complete with black holes. Not to mention, step-by-step instructions for how to take control of your wave and align it to Compassion, with the intention to manifest the greatest good for all. World peace, one person at a time, starting with you.

Dang.

As a high school physics and trigonometry drop-out, I have limited science and math skills. Life with Einstein has taught me how to appreciate Brian Greene,[3] pioneer of super-string theory and author of *The Elegant Universe*. Our maps explain in detail Greene's theory of 13 dimensions. And Nassim Haramein,[4] Director of Research at the Hawaii Institute for Unified Physics. His team of scientists study his unified field theory in order to advance peace and prosperity around the globe. So, too, my Imagined Einstein has an army of Conflict REVOLUTIONARIES®, amateur scientists around the world experimenting with Conflict REVOLUTION® in their own lives, with the intention to become the change. John Hagelin[5] is another renowned peace activist and professor of physics at Maharishi University of Management in Iowa. All three have developed unified field theories that greatly parallel our Einsteinian work.

My challenge is, I am not a scientist. I do not understand mathematical equations much more complex than $E=MC^2$. I would never claim to be able to tell Dr. Brian Greene about super-string theory. All I ever wanted to be was a rock star. I got lured into this divine experiment, I suppose, because I'm good at articulating. I have often said of my relationship to my Imagined Einstein, "I am just the steno."

Nor do I believe my Imagined Einstein means to deliver mathematical equations for how matter manifests. Our work is about telling the story in practical, non-scientific terms in order to teach ordinary people about their extraordinary, inherent power. Others who excel in mathematical equations will hopefully be inspired to articulate our theories thusly. Now that LIGO has confirmed the presence of our gravitational waves, we are well on our way.

---

3 http://www.briangreene.org/
4 http://resonance.is/explore/nassim-haramein/
5 http://www.hagelin.org/

Come with me now on an intricate, intimate journey into how matter manifests and how your relationship to your three Human Dimensions influences the manifestation of that matter. My hope is to inspire you to new ways of thinking and acting to free you from the degenerative dramas of conflict and redirect your precious life force into becoming the change you wish to see in the world, first from within. In this way, one person at a time, we change the entire world.

Part I: The Science of Compassion is the scientific theory behind why Conflict REVOLUTION® succeeds as it does. These technical descriptions and maps explain how matter manifests out of nothing and builds on itself until it becomes your physical reality. Part I is complex; it is meant to inform science. Don't be daunted. While we urge the non-scientifically inclined readers to study the Science of Compassion, it is not required to learn Conflict REVOLUTION®.

Part II: Conflict REVOLUTION® Learning Project is the step-by-step instructions for how to break down a real-life conflict and create a new plan to influence the outcome of that conflict for the good of everyone involved.

Part III: The Action Plan is the self-instruction and self-inspiration to continue your own experiments in this quantum level of self-awareness and personal responsibility. Here you make an active commitment to being an everyday peacemaker.

Do you long to live a magical, creative, and rewarding life, no matter what your circumstances? Do you also long to make a difference in the world today? It's not only possible, it is highly probable that if you are reading these words, you are already on your way.

Introduction

But don't just read the words. Use them to learn to live intimately with your Human Dimensions. Have the courage to apply the process to your conflicts. Take control of your Domain, align to Compassion in the form of *consciousness to the third power* (Cn^3), and empirically experience the miracles that will be created.

The by-product becomes world peace, one person at a time, starting with you. Then, as Einstein, et al. are so fond of saying, "Watch and be amazed."

Barbara With, 2016, Madeline Island, Wisconsin

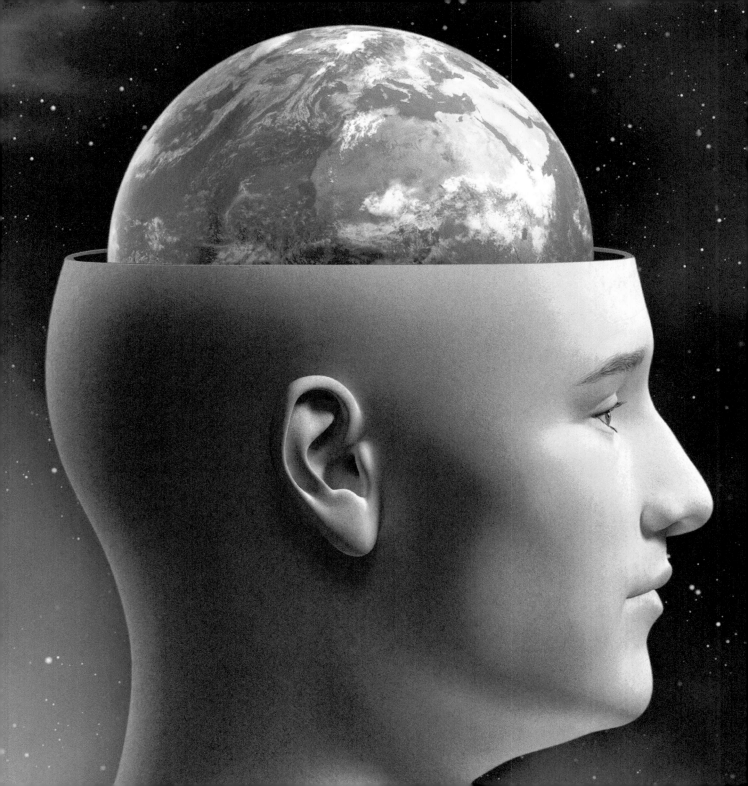

Prelude

The New Operating System

Tombstone of Yolanda Gigliotti
Montmatre Cemetery, Paris
November 2014
Photo: Barbara With

Prelude: The New Operating System

The Party

This journey you are going to take into the heart and breadth and depth of how matter manifests is absolutely and unequivocally connected to you: your thoughts, your decisions, your perception, your emotions, your intuition, your body. You, after all, are the source of your perception of reality. Your body is the gateway; your thoughts, feelings, and senses work through the body to perceive the very human experience that your mathematics are programmed to create.

Gaining mastery and control over your own Domain and consciously aligning to Compassion creates a passionate, present-moment relationship to the now, opens the path to health, fulfillment, and vitality, and contributes to the macrocosm of world peace.

The overarching vision of the new operating system is that the only authority you will answer to is that of Compassion, experienced as your own Intuition, impelling from within. When your three Human Dimensions are aligned to Compassion in the form of *consciousness to the third power* (Cn^3), you become so emotionally fluid and intellectually clear that when Intuition sings out, Her melody inspires your feelings to wed with thoughts that lead to decisions that implement Her calling. To bring the Intellect into alignment with Intuition while feeling and breathing Emotion is the end goal for all that you wish to manifest. This is Cn^3.

When your three Human Dimensions are in such alignment, you are in a position to manifest the greatest good possible in the physical world. This posture is done for the good of the whole, as well as to assist in making what we call "all your wildest dreams come true."

The journey for planetary peace is the same for all: you as individuals must go within and find your power, identify your imbalances in order to correct them, and then take steps to restore balance using simple but effective perspective changes. You will watch and be amazed at how much power you truly possess when you take control of your own Domain in this way.

This journey is an intricate, intimate study of self. You are your own guinea pig, learning to understand how this personal power can work for you and the good of the whole. Is this an unrealistic expectation to believe you can influence the manifestation of physical matter? No, it is not unrealistic, and yes, absolutely, you have the potential to learn how to do just that.

Access to this power, however, will not happen without great commitment to yourself and a demonstrated alignment to Compassion. You cannot do it unless you are ready to face your darkest self, your biggest ego, your most sorrowful pain, and your most unforgivable sin. To deny these parts of self prohibits unity. To bring unity to all your aspects, you need to learn exactly who it is you are. Not who you *think* you are, but who you truly are by witnessing your thoughts, feelings, senses, and the decisions you make. This is a journey of 100% responsibility—full disclosure, steadfast intention, courageous self-scrutiny, and unwavering commitment to self-awareness.

What we can tell you, without a doubt, is that the creation of your world begins within you. Your perception originates in your body. You have the potential to take control of your own Domain and contribute to the regeneration of the planet, as well as creating that life you long to live. You can learn what you need and how to manifest the resources to meet your needs, all the while contributing to world peace by working for the good of the whole. You literally become the change you wish to see happening in the world around you.

Do you have the potential to create that kind of life for yourself? Yes. Can we guarantee an outcome for you? Can we predict if you will make the choice to pick up your power and do the work that only you can do? No. That we can neither predict nor guarantee. You and only you have control of your Domain.

You have the potential to create exactly what you need, but only you decide how much you will stretch yourself past your current condition to learn to have control over your own

Prelude: The New Operating System

Domain. Only you can feel your feelings, think your thoughts, sense your intuitions, and then make your decisions. You decide how to inspire yourself. You create the matrix that will guide your energy into alignment.

Are you willing to tackle the obstacles within? Are you willing to face your own ego? Are you willing to let go of the illusion that this physical world is the primary reality? Are you willing to let go of the physical world completely by facing the death of your physical body?

Achieving this ultimate detachment opens the door for unlimited potentials. For it is when you cling to projected outcomes that possibilities become limited.

This line of study is for those of you who want to go deep into your own power to create world peace, as well as experiencing life as big and rich and passionate. This is for those who want to live in the world but not of it. We speak to those ready to manifest true greatness for the good of all. May your visionary example help lead this planet into a new era.

In that way, we are honored to serve you.

Act I

The Science of Compassion

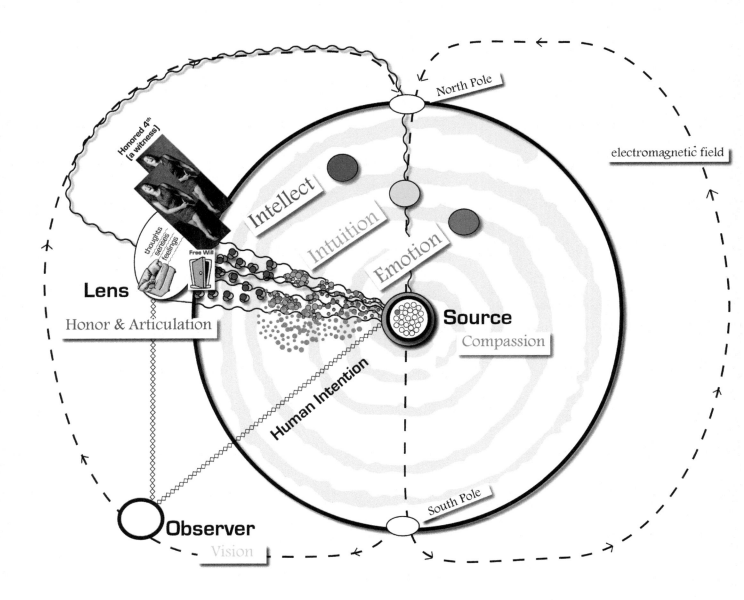

Map of Human Consciousness: Your Domain

Act I: The Science of Compassion

The Origins of the Vision

To begin, let us assure you that this worldview draws on age-old theories and philosophies: Christianity, Buddhism, Judaism, Mayan principles, shamanic persuasions, the Essences, Atlantis, the Aborigines, the Sámi, the Inuits, the Mu, et al. Using the wisdom extracted from these diverse spiritual cultures, we have written a recipe for manifestation of world peace, one person at a time, starting within each individual.

This process is also drawn from the sciences of the Earth—geology, geography, biology, cosmology, etc.—as well as the sciences of the physical body—neurology, physiology, psychology, etc.—to create what we feel is the most comprehensive view of "everything," aka a unified field theory. Our proposal brings together all energy into one creation—you—experiencing everything there is to feel, think, and sense of the physical and energetic worlds.

The bottom line is that our process of manifesting peace cannot succeed unless you commit to becoming a conscious channel for the newly-discovered fifth fundamental force of the universe—Compassion—in the quantifiable ways we will describe.

Making decisions purely from Intellect alone, without considering the impelling of Intuition, creates limiting thoughts that automatically marry themselves to deep Emotion that then get projected into the Lens as imbalanced physical forms. These imbalances appear to be beyond your control—draining dramas, health issues, dysfunctional relationships, mental suffering, etc. However, the powerful images of these manifestations revolved 180 degrees directly back to you places control into your Domain, where your power actually lies. Using the details in that mirror image of the physical world, you can take control of your Intellect and change the stories you are telling yourself. That, in turn, allows you to feel and breath Emotion and hear Intuition guiding you back to balance, one baby step at a time. Emotion can now flow and, following Intuition's impelling, Intellect now makes new decisions that also honor Emotion and Intuition. These new, self-driven actions then restore the balance and align the system to Compassion again—for the good of the whole.

This awareness and corrective action are your responsibility. Practicing this depth of self-scrutiny is a form of self-love, as well as the basis of Conflict REVOLUTION®.

As you keep applying the 180-degree revolution, you train your Intellect to take direction from your Intuition, while you feel and breathe the powerful fuel of Emotion into your body without attachment to limited thinking. The shame of your sins becomes the rocket fuel for your dreams. Judgmental, degenerative thoughts are replaced with the superconscious—powerful, positive intentions and passionate visions dreamt in Technicolor. Emotion is now free to fuel those voluntary visions to become physical reality.

Meanwhile, Intellect is rapt with the dance of creation taking place in the now. Listening for Intuition's broadcast of the small, declarative instruction—"Take this next step for the good of the whole situation"—Intellect decides to promptly take it. These few steps free the Abscess of Emotion trapped within you, releasing stress and greatly reducing the fluctuating waves of Intellect. Empowering, nonjudgmental thinking emerges and you become a witness to the moment of creation with an intentional influence upon it.

And all of this, miraculously, begins with nothing.

Starting from zero, we will illustrate, one step at a time, how matter is manifest into your Domain, including:

- The location of the Void and the Source;
- The quantifiable definitions of Compassion and a *compilation of consciousness*;
- The exact nature of the *Human Intention*: the mechanism that organizes energy into physical form, including the *Source, Observer,* and *Lens*;

Act I: The Science of Compassion

- The three Human Dimensions of Emotion, Intuition, and Intellect and how they operate individually and with each other;
- The Honored 4th and how to cultivate a reference frame to witness self; and
- The form and function of the human body.

This is a process for taking complete control of your own Domain. There will be relationships drawn, homework given, and expectations that you will set up for yourself. Your new operating system will be self-designed, self-inspiring, self-regulated, and operating for the good of all the parts of the whole.

You will be asked to remember the times you reached through a challenge and came out the other side with wisdom and compassion. We will use that success to impel you to take another step. Remembering your successes can inspire you to go into the darkness, peel back another layer, and have the courage to take a breath and face the truth of who you are without any illusion. Dedicating yourself to these simple steps every day will create miracles.

Always remember, what you choose to do with your power affects the planet as well. It's a "three-fer"—you, us, them—Cn^3. Harnessing your personal power is done for the greatest good—all of us consciously coming together in the Source to create worldwide peace.

Commit to the Mystery: M-theory

To start, we tell you up front: the root of all creation is a mystery. The meanings of "emptiness," "vacancy," and "vacuum" depict the zero point, where nothing exists. We repeat, the root of all of everything is a giant mystery and always will be.

M-theory is a theory of physics that unifies all other versions of string theory. String theory accords consciousness into strings; M-theory creates three-dimensional circular membrane domains that surround the universe contained therein. According to Edward Witten, the

Act I: The Science of Compassion

theoretical physicist who invented the term,[6] the *M* could stand for "membrane," "mystery," "magic," "manifestation," or even "miracle" and should be used according to taste.

So right from the start, commit to many mysteries. Understand there is much you will never know, so accept it. The more that you can live with the statement, "I don't know ...," the more powerful you'll become. This statement will quiet your Intellect and allow you to relax and witness the creation of your own life.

Mother Earth: There's No Place Like Home

Imagine that Earth has a ring, like Saturn, circling the planet like a halo. Please have a seat on the edge of this halo and imagine yourself a satellite looking down on Mother Earth. From this ledge you can observe the activity going on below. See the formations of all of the terrain: the mountains and the oceans and the continents—blues and purples and greens and browns—all the colors of our beautiful planet. We call this point out in the heavens the *Observer*.

So here we all are, looking down on Earth, a beautiful, glowing blue orb of life. It makes you weep to see this beautiful planet, this oasis in space, alive and holographic. Breathtaking, is it not?

This is heaven. Literally, Earth is the garden of legend. There is no other place in the entire universe that provides everything you need to be human—water to drink, soil to grow food, air to breath, fire to sustain and transform, and communities to feed human hearts. The planet Earth originates your perception, nourishes your body and soul, and is the heaven of your dreams.

As you look down on Earth from the Observer, find the location where you currently are on the planet. We call the place on the surface of the planet where your body is the *Lens*. Using your microscopic vision, see yourself sitting wherever you are in the Lens, reading these words on the page. Like a zoom on a camera, focus in and feel yourself now in your body situated in

---

[6] https://en.wikipedia.org/wiki/Edward_Witten

the chair reading this book. Then move your focus back out in space at the Observer, and then go back again to your body in the Lens. This is like being in two places at once.

Now bring your awareness back out to the Observer and allow us to escort you to a third point, which is really the first point, in the center of the Earth. This is the location of the source of all life as you know it. We call this the *Source*.

The Void: Making Something Out of Nothing

The Source is an unseen singular point of origin inside the center of the planet, from which gravitational waves radiate. The core of Earth is spinning much faster than Earth itself and is much hotter than the surface of the sun. This energy is so powerful that spacetime warps, creating the conditions of a singularity at the bottom of a black hole. Gravitational waves result from the constant Big Bangs taking place as the black hole separates and then explodes back into the singularity. These waves are the medium on which physical matter is then constructed.

All of the particles traveling up the wave are reflections of the one singularity of the Source. The wave, like a river of floating leaves, carries infinite reflections of the one singularity, each with its own unique reference frame. This is what allows the singular point of origin to be perceived as infinite multiples by the emerging consciousness.

The simultaneous presence of pure energy and absence of all light where the singularity resides is the *Void*. Some call it zero point. If you could stand on the event horizon of this Void, you would see that the heat and light of the core are so intense that the point of origin is totally inaccessible to direct observation. Conversely, since there is no light within a black hole, you can never see the point of origin, although you can measure the gravitational waves emitting from it. These conditions allow for infinite mystery and create a perfect paradox: everything and nothing, light and dark, silence and sound, separation and unity, all together in the same consciousness.

Act I: The Science of Compassion

Mathematically, the first singular point does not allow for self-awareness. To be the singular point, there can be no other point. There is no observation of consciousness at the point of origin, but that does not mean there is no consciousness. The observation of the Source, however, is as essential to the creation of matter as the Source itself. This addresses the question Einstein posed in his life about looking at the moon. If you are not looking, he thought, surely the moon is still there. It is there even without anyone observing it, but without the observation, the moon is just another photon of information filled with mathematics until it is observed.

The singular point of origin has only one possible reference frame, but the gravitational wave emanating from it provides the opportunity for perceiving infinite multiples. And because your consciousness is a part of both the singularity and the gravitational wave, you can achieve a reference frame from anywhere on or outside the wave. This makes you the Source, the Observer of the Source, and the Lens from which to perceive the Source.

Yes, the center of the planet is the geographic location of the Void and the Source, which is the beginning of all human life. At that location are contained all of the non-physical mathematics that are at the root of all the physical matter within the Earth system that you are perceiving.

The center of the planet is estimated to be anywhere from 2,000 to 7,000 degrees Celsius, about the same temperature as the surface of the sun. We believe, however, that it is actually much hotter. The deeply condensed core of iron spins with such force that it nearly bores a hole right through the fabric of spacetime, creating a condition like the void of a black hole. The heat, however, holds information. Here the shadow of the physical world is cast in what some might call "anti-matter," the root of the projection that will become the matter: the mathematical equations of the who, what, where, when, how, and why of manifestation.

All combinations of potentials exist in the Void and the heat of the molten, spinning core of Earth. From this paradox of all and nothing, from the heart of zero arises the first step into physical matter: one.

Scientists today work within a system comprised of four fundamental forces:

Electromagnetism: The force associated with the behavior and interaction of electric and magnetic fields, such as electricity, magnetism, and all forms of electromagnetic radiation, including light. Electromagnetism is responsible for igniting information.

Gravity: The force of attraction that all objects with mass have for each other. At the quantum level, where mass is very small, the force of gravity is instrumental in pulling the compilations of consciousness into their given form. Gravity is responsible for guidance.

Strong Force: The force that mediates interaction between particles. Strong force is responsible for binding and breaking.

Weak Force: The force that transforms the flavor of particles. Weak force is responsible for transforming the nature of one thing into another, e.g., from zero to one, from red to green. It also instigates decay.

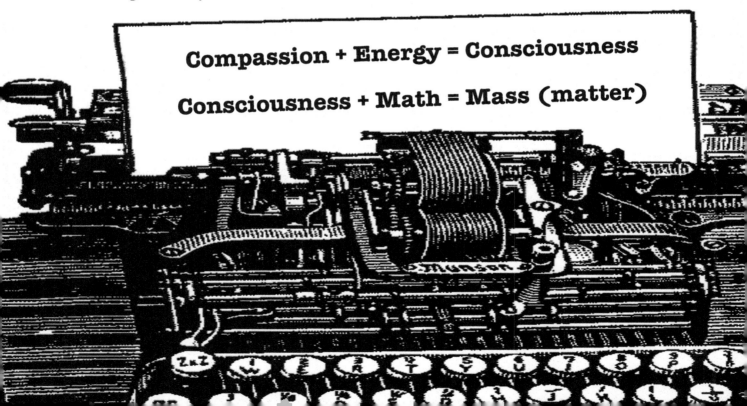

Compassion + Energy = Consciousness

Consciousness + Math = Mass (matter)

Act I: The Science of Compassion

Compassion: The Fifth Fundamental Force of the Universe

What is it that impels energy to take the first step out of the zero and into one to form consciousness? It's as if from that mysterious nothingness comes an urge to create. This insatiable longing, this passionate pulse of intention to create is a mysterious stirring and the power of the fifth fundamental force of the universe, Compassion.

Compassion is the intelligence that uses the four fundamental forces to impel creation of the physical world one step at a time.

When Compassion fuses itself to energy in the Void, consciousness is created. When consciousness fuses itself to the math in the Source, matter is created.

Utilizing the four fundamental forces, Compassion influences energy to consciously behave in three specific ways:

Circles. Working like a compass—the instrument for drawing circles and measuring distances—Compassion impels energy to form spherical electromagnetic membrane domains that encapsulate certain subsets of the mathematics in the Source that define the specific potential to be manifest as and through you. We call this spherical Domain your *compilation of consciousness*—an electromagnetic particle containing the mathematics of your identity as well as the identities of all of the physical manifestations of your world.

You are all rooted literally in the center of the planet Earth. You all share the same math in order to uniformly perceive the physical world—the chair, the trees, the entire planet and, in fact, the whole universe. This makes you a subset of Earth, as well as one with one another.

This compilation of consciousness is the beginning of your Domain.

True North. A compilation has exactly the same energetic structure as that of planet Earth. Compassion impels each compilation to align its electromagnetic field to the magnetic north of Earth, like a compass—the instrument that determines directions utilizing a freely rotating magnetized needle. (See page 20.) This creates the key for the map of the location of where and when physical manifestation will take place in any given moment. Latitude, longitude, form, time, color, vibration—all of these mathematical formulas eventually manifest on the surface of the Earth as a perceived physical object with shape, location, and attributes. Compassion ensures everyone is using the same map with the same features and attributes. The only thing that differs is perception. While you can all see the chair, no one else has the exact math to create your reference frame of the chair but you.

Conscious Connection. Compassion aligns your compilation to 0 degrees, making you zero degrees separate from Compassion. This in turn creates communication between and awareness of the parts of the whole.

All humans have their own Domain—your mother, your daughter, your teachers, your cousins, every stranger, et al.—but all compilations intersect in the center of the planet in the Source at 0 degrees as one organism. Because of this, the compilations communicate with each other, working in concert to eventually create the whole of the physical universe.

Imagine the center of the planet is filled with compilations that contain all the mathematics of all of the humans who are having or have ever had the Earth experience, all packed together in the wildly vibrating world of quantum. Compassion aligns them to true north and binds them into relationships each with one another in preparation for creation.

The compilations in the Source do not touch one another. Their electromagnetic fields keep them separated. As they interact, however, the friction creates sparks that literally explode between them. Acting like a Big Bang, as it were, the sparks ignite the next step into physical

Act I: The Science of Compassion

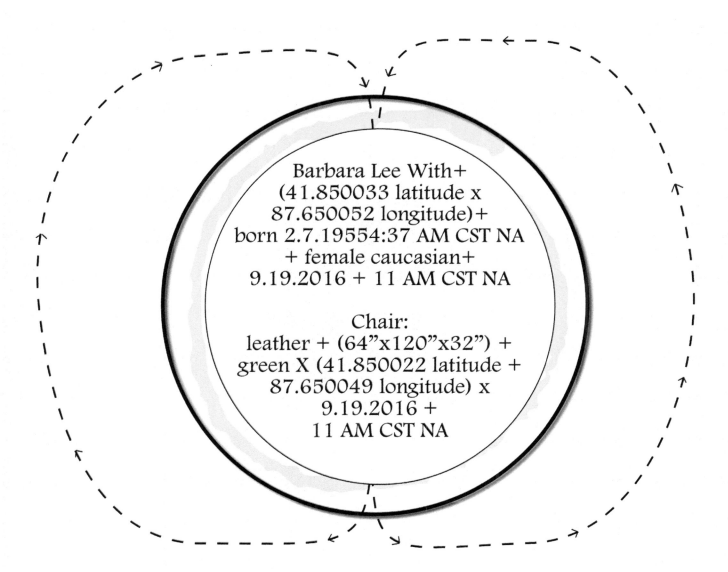

A Compilation of Consciousness

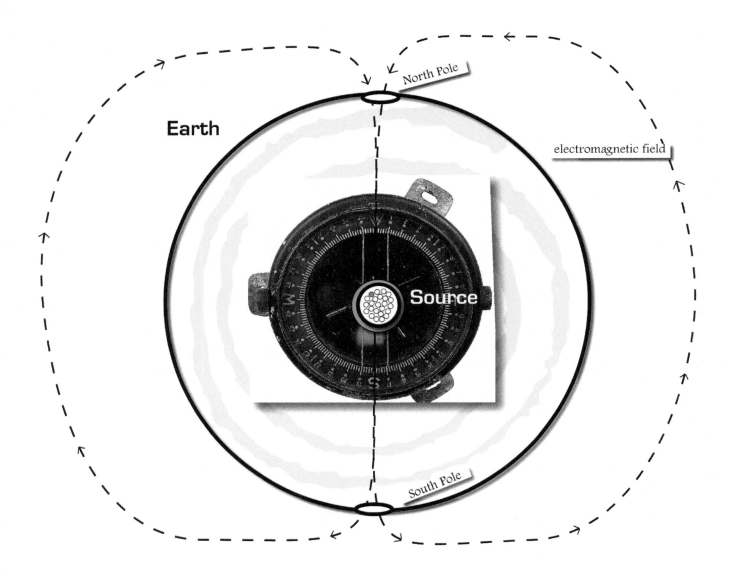

Compassion Aligns the Compilations of Consciousness to 0° True North

Act I: The Science of Compassion

form: the beginning of your string—a gravitational wave flowing from the Source that eventually evolves into the human experience of the physical world in the Lens.

When this wave reaches the surface of the planet and bursts into the outer world of Earth, the Lens of physical perception is created. Here energy slows down and separates into your physical body, as well as the world your physical body is perceiving. While a certain part of your energy gets "trapped" in the Lens as perceivable mass (your body), the wave continues onward into space, out to the edge of the Earth's electromagnetic field. From there, it flows back up through the North Pole, down into the center of the planet, and comes back to the zero point of the Void to begin again. This eternal circle of life is taking place at the speed of light squared (C^2) and is the string theory about which scientists speculate.

Every physical object in the Lens has its own unique vibrational frequency. These vibrations are the individual voices in the great song of the entire universe being sung into creation. In the Bible, sound came before light. "And God *said*, 'Let there be light.'" Before light, which is carrying information, there is sound, which is impelling action. And before sound, there is desire, which is the mysterious urge to create that consciousness has at its root: Compassion.

Your string, or gravitational wave, is filled with reflections of your compilation that arise from the Void into the Source. The reflections are all vibrating at the same unique frequency that is you, the human in the Lens who is reading these words. An inner gravity of desire is pulling together all the reflections of you into the form of you. Another name for this vibrational fingerprint is *sexual energy*, which plays an essential role in building the physical world.

This gravitational wave that is you is always on a journey into perfect manifestation: whatever is programmed into the mathematics at the level of the Source will perfectly manifest into physical form in the Lens on the surface of the planet. Those physical manifestations are also influenced by the condition of the relationship between your three Human Dimensions. If there is conflict between Emotion, Intuition, and/or Intellect within, that conflict will manifest perfectly in the physical world.

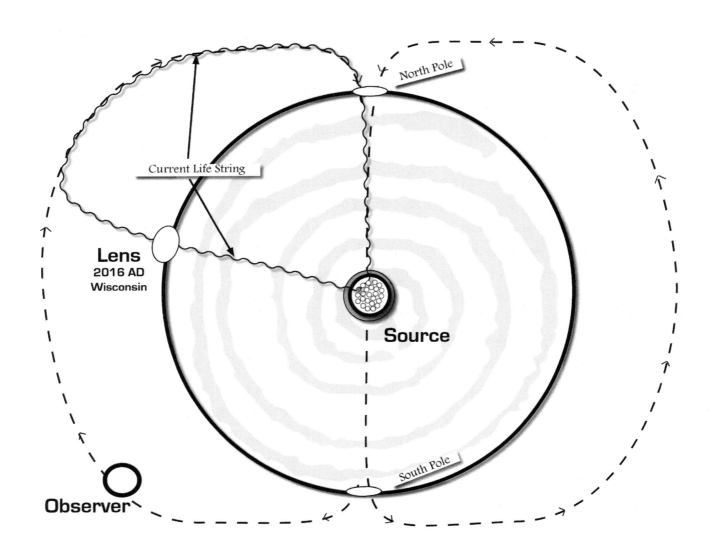

The Human String Theory

This is why everything is "perfect" no matter what manifests in the Lens. The system is perfectly manifesting the degenerative mathematics of this inner disunity into conflicts, destruction, illness, and chaos in the Lens. It is, indeed, a perfect system, but it's not being used for the greatest good.

In order for you humans to survive the seven billion and growing, you must reprogram your operating systems, one person at a time, to process human energy in new, regenerative ways. Your individual gravitational wave is where the work to build a new operating system will take place. The challenge is that Intellect does not have a natural capacity to make sense of the reality of this subatomic level. You can, however, examine the manifestation of the wave in the Lens and the details of how the conflict is manifesting there to understand the conflict on the wave itself. Thus, a revolution of your perspective will be required.

For this, we'll put you right back into your physical body, in that chair, reading these words right now on the surface of the planet. By examining what is being manifest there in the Lens, you will find the guidance as to what and how to change the inner quantum level of your operating system. Whether you choose to make those changes, however, is entirely up to you.

The view from the Observer will play an important part, as will the view from the Source. You will learn how these three points interact to guide this gravitational wave to create the perception with which you experience the physical world and how to have control over it. The condition of this wave flowing through your body at any given moment will become your primary concern and is the subject of your Domain.

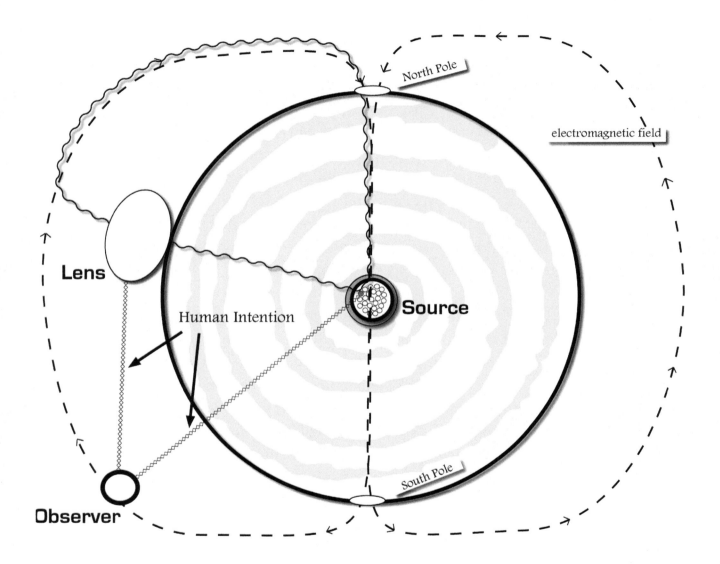

The Human Intention: Source, Observer, Lens

The Human Intention

The three locations where we've placed you—the Source, at the center of the planet; the Observer, out on the edge of the Earth's electromagnetic field; and the Lens, cemented into your physical body on the Earth encapsulated in form—are three parts of a living, breathing mechanism that allows the spinning string of energy to slow down and separate, pulling together into physical form as perceived through your physical human body. We call this mechanism the *Human Intention*.

Your gravitational wave is like a spinning string that contains the whole of your human experience. But what brings your string to life is its intersection with the Human Intention.

The Human Intention is a triangular energy structure consisting of these three points—Source, Observer, and Lens—that operates like a prism, providing substance and separation to the light and sound particles on your gravitational wave. Human Intention also reveals the three Human Dimensions—Emotion, Intuition, and Intellect—and creates a free-will body with thoughts, feelings, and senses to be alive in the Lens of the physical world.

Your body is being created as per the math in your compilation, including the spacetime coordinates of where you are on the map and the DNA of the actual physical form that you inhabit. Compassion creates communication between the parts of the whole, allowing humans to be self-conscious. You have the ability to analyze the spectrums of the three Human Dimensions. This is the "human" intention of your experience, as opposed to that of the chair, tree, or star. Only human beings possess the power to look at and be aware of self from their physical reference frame and have the power to exert influence over that physical world.

Source: The origination location of all consciousness in the form of compilations, located in the center of the Earth. All compilations originate in the Source, which originates in the Void. Source is associated with Compassion because the Source is zero degrees of separation from Compassion. Emotion is the Human Dimension associated with Source, Compassion is the Human Affiliation, and feelings are the body's direct connection to Source.

Observer: A point of consciousness on the edge of the Earth's electromagnetic field that becomes the fulcrum between the Source and the Lens. Communicating with both the Source and the Lens, Observer reads the present-moment condition of the physical manifestations in the Lens and relays that information back to the Source at the speed of light squared (C^2). That information then becomes a part of the mathematics of the wave making its way to the surface of the planet to become you. This is what allows Intuition to know what direction to impel for the good of the whole. Intuition is the Human Dimension associated with Observer, Vision is the Human Affiliation, and senses are the body's direct connection to Observer.

Lens: The location of the physical manifestation of your compilation of consciousness as per the mathematics contained therein, as well as the perception point of the physical manifestation. Your body is manifest from the inside out and is the vehicle for the projection and perception of the physical world. Intellect is the Human Dimension associated with the Lens, Honor & Articulation is the Human Affiliation, and thoughts are your body's direct connection to the Lens.

So you are not just a compilation; you are also a spinning string and Human Intention—the sacred geometry of the triad, the holy three. With three dimensions you have all space; with the three primary colors you have all colors; with light, sound and color you have physical reality;

Act I: The Science of Compassion

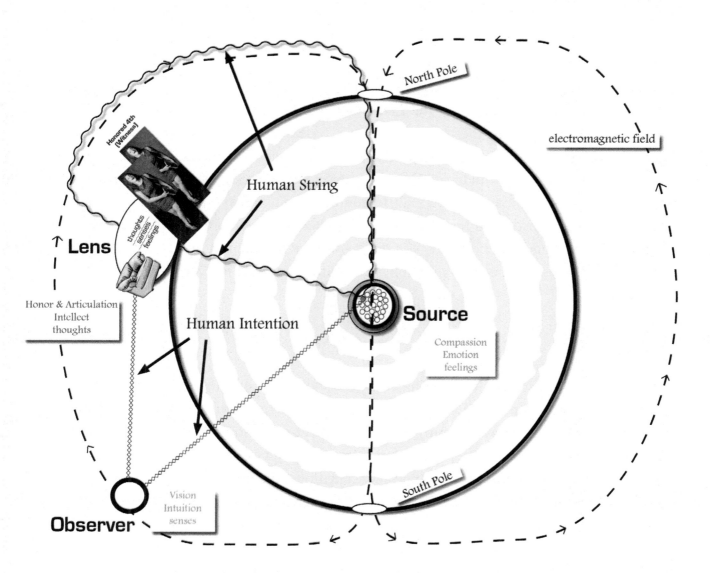

Human Intention Interacting with the Human String

with the three Human Affiliations you have a psyche living in that physical reality; and with the three Human Dimensions, the psyche has the ability to perceive and analyze the infinite glow of physical reality.

When your physical body dies, the Lens retracts back into the Observer to be refocused on a new and different manifestation experience based on different mathematics. Perhaps the new configuration of mathematics will take you to Ireland in 900 AD in a man's body. Different DNA, different latitude and longitude, different time codes, same consciousness. But the string that was and had been you in 2016 continues spinning as a hologram in the form of thoughts, feelings, senses, and memories of the life that you lived in that body and spacetime. Without Human Intention, however, you don't have the mechanism to slow down the string to organize and separate the energy into physical matter. You spin forever, shining like a star in the heavens.

$$E = MC^2$$
$$E \div C^2 = M$$

The E (energy) of your string is spinning at C^2 (the speed of light squared). At the point of the Lens, speed slows down to the speed of light (C) and energy is divided into separate form, creating M (the mass of physical form). The Lens, like a prism, slows the flow of the sound and light particles on the string and separates them into colors and tones, form and function, as well as provides the platform from which to perceive creation—the human body.

After passing through the Lens, the wave accelerates again to C^2, carrying with it the records of all the real-time decisions that are made in the Lens. If you choose, for example, to act in anger and rage, a particular set of circumstances will take place. If you choose to act in love, a different set of circumstances will transpire. These "If P, then Q" formulas become a part of the mathematics of your compilation as your string speeds out into the heavens to join up with the

Earth's electromagnetic field and get swept up through the North Pole and back into the Void again. This "compiling" of your decisions becomes a part of your math on the next spin up the wave to become karma. In this way, present-moment decisions influence what will manifest in the Lens in terms of the "future." You literally reap what you sow in this karmic way: if you react in rage and anger, it revolves 180 degrees the next time around to get refocused directly back at you.

When the physical body dies, your gravitational wave accelerates back to C^2 as the reduction of M leads to an increase in C.

This is how we explain Afterlife and how you can perceive the Party's presence: our consciousness in the form of a hologram of thoughts, feelings, senses, and memories still exists even though our physical bodies do not. We haven't gone anywhere; there's literally nowhere else to go. We're still spinning. It's eternal. The Lens doesn't trap us to slow us down into separation; we are pure energy. With her permission, we are borrowing Barbara's Lens to communicate. We are riding her wave with her, sharing her three Human Dimensions. And it's a wild ride!

Where All Your Power Lies

Your personal power resides in coupling the three Human Dimensions that comprise your gravitational wave with your self-awareness. We will teach you to use the power of self-awareness to thoroughly study the nature, condition, and function of these Human Dimensions within you so you can make conscious choices to exercise your power over them to do what is best for the good of the whole. To do this, you must understand the individual purpose of each dimension, how they influence one another, the condition of your own Human Dimensions, and how to take action to change what is in conflict at this level.

So what are these three Human Dimensions? What are their individual functions and how do they work together? What condition are they in within your own Domain? What does it mean to say that they are in conflict? What can be done to resolve it?

How can you become aware of how these Human Dimensions are functioning in present moment? And how can that reality become your primary reality and the physical world become the secondary reality?

In our next chapter you will take an in-depth look at the three Human Dimensions. As you go deeply into each and understand them from both a quantum as well as a physical perspective, you learn how to have power over them.

Homework

Be prepared to pay attention and watch yourself—what you are thinking, feeling, and/or sensing in any given moment. Sometimes it's easier in the "unimportant" times to watch because you don't *have* to be thinking. Standing in line in the super market does not require any particular thoughts. Paying for your groceries and interacting with the cashier does. Pay attention to both.

Imagine this sensation of being a string of speeding light, color, and sound, circling the planet, intersected by the three-part prism of the Human Intention. Put yourself in each place on the map, not necessarily just when you're meditating, but when you are out in the world, too—in line at the supermarket, or in the dentist's office. Create a game for yourself: instead of worrying about yesterday and tomorrow in those "unimportant" moments, focus your perspective on the present moment. Focus on the bigger you in the Lens, then out at the Observer, and then down in the Source. In experiencing the parts of the entire system individually, you discover that the physical world is only a small piece of an infinite, mysterious operating system and you are the center of your universe.

Notes

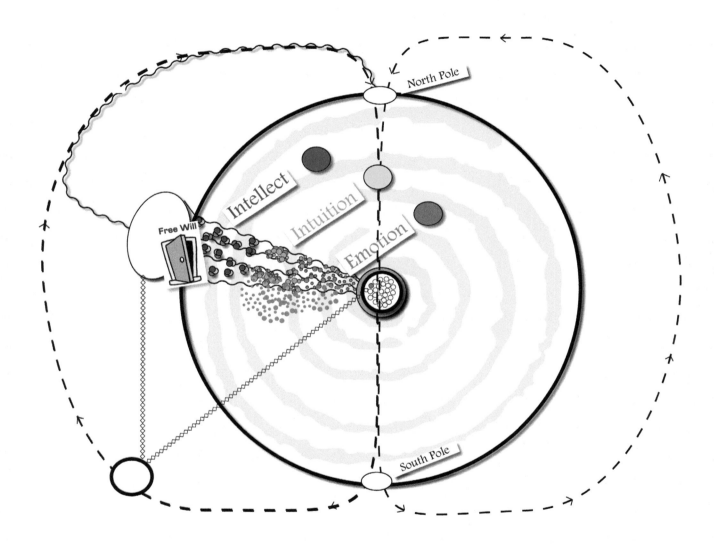

Three Human Dimensions: Emotion, Intuition and Intellect

The Three Human Dimensions

Begin now by adopting the perspective of Observer. Look down on Mother Earth and see the terrain—land masses and waters—and all the colors of our beautiful home planet. Take a long breath and drink in the vibrant glow of heaven. (See page 12.)

Using your microscopic vision and vivid imagination, look down to where your body is physically situated right now in the Lens. From afar, observe yourself from the *outside in*: you, the chair, and the book in your hands.

Now go into the Lens and put yourself into your physical body. Notice the room and the chair from that perspective—the *inside outward*. Feel the clothing you are wearing against your skin, this book in your hands. Feel this separated reality using your five senses of touch, taste, smell, hearing, and sight. Notice that you have the ability to witness both the inner, non-physical world of Emotion, Intuition, and Intellect and the outer world of objects.

First Point

Descend deep into the center of the planet to the Source and find the compilation that is you, emerging from the Void to surround your subset of mathematics with an electromagnetic boundary. This is the first point in your infinite world. Like a personal spaceship enclosing your consciousness, your compilation buzzes with the hum of your own electromagnetic field. It is represented by the lone red dot in the center of the Earth in the drawing on the previous page.

Eventually, everything contained within the compilation will be projected outward to become physical reality in the Lens. What is humming is the living information contained within the compilation that is in a constant state of creation. At your root, you are a unique song,

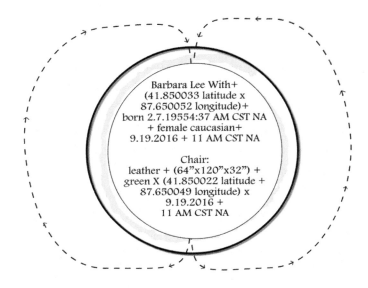

First Point: The Original Compilation

as is the chair. What brings it all to life is Compassion impelling electromagnetism to ignite the information into a wave that becomes your body and the physical world. The Lens is always a perfect projection of whatever the mathematics are at the Source.

This buzzing of the math is also a kind of inner gravity that will be pulling together all the reflections of your compilation as they are moving up the wave toward the surface of the planet. This sexual energy is an essential part of the process of the creation of your unique physical body.

Compassion aligns everyone's compilations to true north. This way, every human has the same map so you can find your way around the Lens together.

Second Point

After the creation of the one compilation, Compassion impels the weak force to split the one into two. The creation of the second point is the next step to building a dimensional form. This separation creates the inner world of you and the outer physical world that you will perceive.

These two points are in constant communication with each other. Both are still contained in the compilation, but they are morphing, like tiny black holes within your body, dividing and multiplying. This splitting also serves like a Big Bang, sending another gravitational wave moving up to the surface of the Earth towards the Lens.

Act I: The Science of Compassion

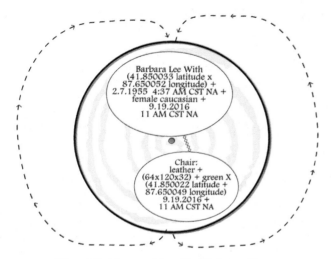

The Mathematics Split Into Two

Even though there are now two points, consciousness can only perceive one. The reference frame of point 1 is of point 2—only one point. The reference frame from point 2 is point 1—still only one point.

Third Point and Infinity

The magic number that sets the mechanism spinning into creating the miraculous, infinite, and complex system that becomes your physical experience is the arrival of the third point. The reference frame from this third point allows for self-awareness of both in the inner and outer worlds, as well as the entire compilation.

Now when point 1 looks out, it sees point 2 *and* point 3. This sets up the potential to experience infinite multiples. Perceiving two at once creates the potential to see three, four, five, six, seven at once, on into infinity.

When point 3 arises, another wave explodes from the compilation and the Lens drops into place to slow down the spinning string, like a prism separating white light into a rainbow of unique colors. This allows for the perception of separation as the human body is created and, using thoughts, feelings,

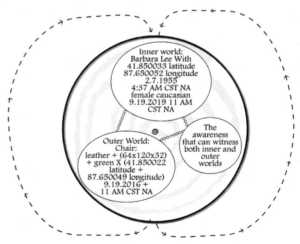

Third Point Emerges

and senses, projects the mathematics of the universe onto the inside curvature of the Lens to experience the separation of the physical world.

Now take a big, long breath. As you feel the air filling your chest in the Lens, see the entire planet from the Observer. From the Source, prepare to ride with the reflections of your compilation along their own gravitational wave, making their way to the surface of the planet to create physical reality in the Lens.

By the time you reach the Lens on your journey into separation, your consciousness will be discernible from the oneness by way of your physical body. Even though you will appear as separate from the chair, your mother, and the man at the gas station, in the Source you are still one with everything you are experiencing in the physical world. You and the chair are literally superimposed on each other in the Source. In the Lens, however, your body becomes the complicated intersection of energy systems that allows for perception of the separation.

Notes

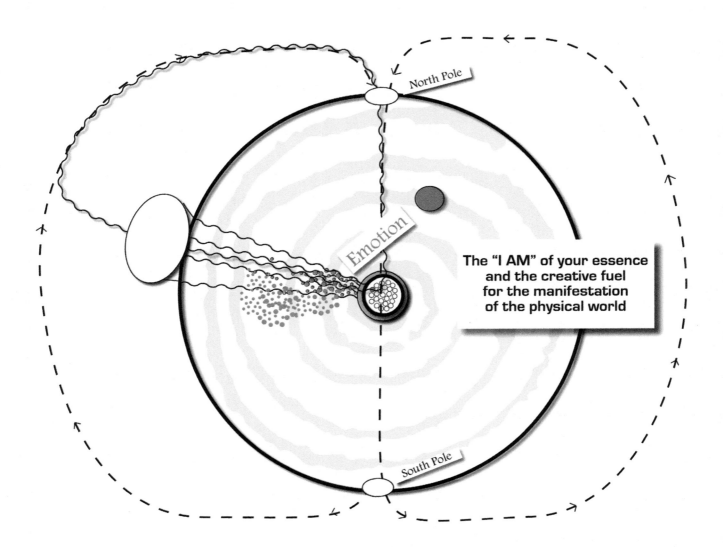

First Human Dimension: Emotion

First Human Dimension: Emotion

Let's examine your wave as it steps out of the Source and into the first leg of the journey into the physical. At the level of Emotion, the reflections of your compilation are scattered and thrown, hither and yon, carpeting the wave, mixing among the reflections of the compilations of all the other physical manifestations that will become your world. At this level, you are in the preverbal primordial soup.

Imagine you have shrunk and are walking along the wave amongst the reflections of the compilations of consciousness. To your left is one of the compilations destined to become your chair. To your right is one of the compilations destined to become your desk. In front of you are several that are destined to become the trees outside your window. Everywhere there are orbs of glowing math, all buzzing and mixed together in this magnificent primordial soup.

At this level, the reflections of your compilation are communicating with one other. Compassion is impelling gravity to guide them to each other. This is the prelude of pulling the "I AM" of you and your math into physical form. This ever-changing I AM math includes DNA, latitude, longitude, present-moment height, present-moment weight, birth date, current date, current time, etc. Your sexual energy is the unique gravity of each reflection, attracting others of the same frequency together like a family. By the time they reach the Lens, they will have all pulled together to create the contours of your body and the forms of everything that you are experiencing around you.

This close to Source in Emotion, the reflections of the compilations of the physical world are all taking the first step towards attraction. They are communicating, like a little roll call. "I

AM Barbara With!" sings one; "I AM Barbara With!" sings another as they begin to make their way towards each other. "I AM the chair!" sings yet another. These are the many voices of the one song of life—the "uni-verse."

Again imagine walking around the compilations on this part of the wave that is Emotion. The chair, the desk, the trees, the window, etc.—everything in the physical world begins inside you. The primordial wave of Emotion is where self-awareness takes its first step. It's not that there's a chair there and you see it. You project the dimensions and the contours of the chair into physical form in the Lens because you are having an intimate relationship with that chair at this level.

This truly is a co-creative process, because without you observing it from your unique angle, the chair is merely mathematical particles riding a gravitational wave.

The Gravity of Sexual Energy

***Sexual energy* is the unique vibration of your compilation, the inner gravity that pulls you together into form.** This more rudimental purpose for sexual energy is often misunderstood in the Lens, where sexual energy has become about pleasure and gratification.

When you are close to someone in the Lens sexually, you are activating consciousness deep inside the planet. Of course, your body can never fully step into someone else's physical body to be in the exact coordinates that they are. However, if you were to take a walk through Emotion on your gravitational wave while you were having sex with someone, you would see two electromagnetic particles vibrating towards each other, eventually overlapping to share the same vibration. You and your partner would be intersecting and begin to take on each other's vibration, pulling you even closer. As your physical bodies mesh in the Lens with breath and movement, your speeding heart rates accelerate all other systems. You two are creating a third new vibration that is greater than the sum of your parts. This is why love affairs and sexual

Act I: The Science of Compassion

encounters can be so energetically powerful: you are activating a new energy that is a unique combination of the two vibrational rates. The eventual orgasm is an expression of a birth that can become the gestation of an entirely new physical life. This is as close to sharing the same latitude, longitude, and reference frame as is humanly possible.

Emotion is the creative fuel for manifestation, the I AM essence of everything physical, and is associated with Source as well as the solar plexus region of the body. Flowing through your physical body, it shows up as feelings that, along with senses and thoughts, create your physical experience in the present moment of the Lens.

On the part of the wave that is Emotion, so close to Source, there is no language or Intellect; no comparisons, no stories, no judgments, just the present-moment experience of the pure I AM of wherever and whoever you are in any given moment.

To experience your consciousness on that part of the wave, before Intuition, before Intellect, and before form, is a position of great power. It requires commitment to the mystery and the ability to surrender to the experience of just being pure energy.

When Emotion does not flow through the physical body in some healthy way, it becomes trapped on the wave. This abnormal cycle of backward flow is called the *Abscess*. We will take a closer look at the Abscess when we examine Intellect.

Expressing vs. Feeling

Conflict REVOLUTION® teaches a revolutionary new relationship to Emotion. In Con Rev, there are no "bad" feelings. All are part of the fuel that instigates the manifestation process. All are as necessary to the creation of physical form as red, yellow, and blue are to the color wheel.

This new relationship to Emotion requires two different skill sets:

Expressing Emotion is an intellectual activity that defines and articulates your present-moment perception of your emotional condition.

Feeling Emotion is a physical activity using breath and Compassionate Intellect to intentionally move present-moment Emotion through the physical body in a regenerative fashion.

We often mistake these skill sets. We *think* we are feeling so much, when we are, in fact, *thinking* about how we feel and talking about it. Thinking and talking are intellectual activities, which can cause anxiety, frustration, and anger to become stuck in the Abscess. Feeling is a body experience. We intend to teach you the difference and inspire you to process Emotion in new, regenerative ways for the greater good.

Expressing Emotion vs. Feeling Emotion

Expressing Emotion is an activity using Intellect to define, honor, and articulate your present-moment emotional condition.

Feeling Emotion is an activity using breath to intentionally move the energy of a present-moment feeling through the physical body without intellectual interference.

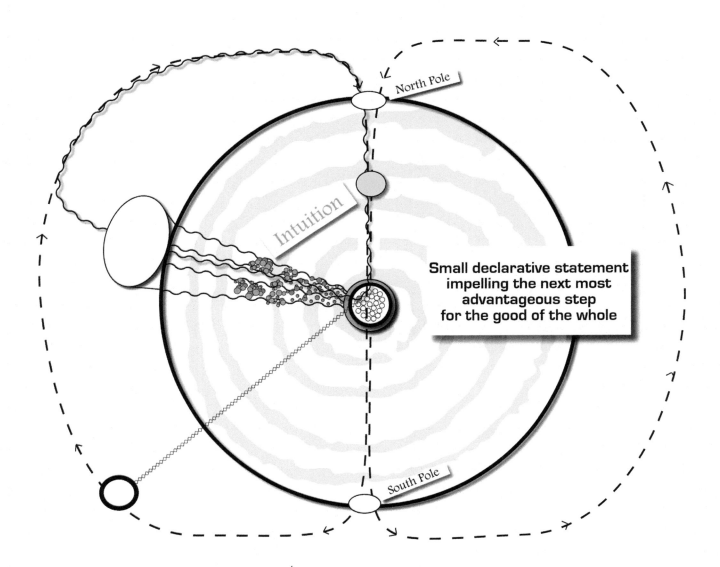

Second Human Dimensions: Intuition

Second Human Dimension: Intuition

Once again, assume the perspective of Observer looking down on the planet. See your own gravitational wave moving from the Source up to the Lens, creating your physical body and filling it up with the math of your consciousness. See that gravitational wave continue on into the heavens and merge with the electromagnetic field of Earth, only to get swept along into the North Pole and back down to the Source to begin again.

Imagine yourself at each of the three positions of the Human Intention: at the Observer, in the Lens, and in the Source.

At the Source, imagine taking one step onto the carpet of your gravitational wave and into the first Human Dimension of Emotion. As you stand there, feel the hum of the universe as the emerging I AM of you and all the objects in your physical world sing the song of creation in this amorphous primordial soup. Immerse yourself in the white noise. Allow thoughts to cease and just be one with everything.

Here on a quantum level you begin to experience separation as a unifying force. To think that unity begins with separation might seem illogical. Consider, however, that when a cave man had a stomach ache thousands of years ago, the witch doctor thought the gods must be angry with him. When medical science emerged, researchers began to understand not only the separate parts of the whole body, but also how they all interrelated. Having an understanding of the separate parts of the body led to a greater understanding of how the whole body worked. This is exactly the same kind of separation that is needed in order to unify the planet on molecular, personal, and global levels.

On a global level, people must strive to accept the unique quality of all cultures and honor diversity. On a personal level, healing of self is about understanding your own specific needs separate from everyone else's needs and finding the resources to meet them. On a quantum level, experiencing the separation of Intellect, Intuition, and Emotion, learning how they work individually and concurrently, and choosing to apply your power to align them to Compassion is needed for the species to survive.

So how does Compassion pull together the reflections of your compilation and move them up the wave to take the next step into becoming matter? Using the weak force, Compassion "transforms the flavor of" the one by splitting it into two. (See page 35.) The strong force then binds those two particles together even tighter into form. The song of your I AM becomes your voice and the awareness of sound is born. Once the one splits into two, out of the unvarying, unobtrusive, and eternal white noise of Emotion, sound can now be heard.

Intuition is the sound of the emerging voice of Compassion impelling the process onward into creation. Here, alto, tenor, soprano, and bass can be detected, and the sounds of the world begin to organize. The song of the bird, the crash of the thunder, the roar of the ocean—all are tuning up at this level of your wave.

Try this: sit where you are and look at the world around you. Whatever it is—the rug, the walls, the chair, the ceiling, etc.—it's being created by color, sound, and light interacting with your five senses to form the physical world. Now, close your eyes and just listen. With eyes closed, sense the world without light, a different orientation than when your eyes are open. When they're open, the focus is drawn to what the light is revealing. When they're closed, you're drawn to what the sound is revealing. Remember, "And God *said*, 'Let there be light.'"

Hers and His

Out of the duality being created at this level also spring feminine and masculine. Intuition, sometimes referred to as the Divine Feminine, is responsible for impelling the reproduction of particles in a way that fulfills the mathematics of your compilation. Intuition knows the complete plan of creation, as well as the next most intricate, intimate baby step needed to keep moving you forward into the physical as per the plan. The feminine emerges first, and from Intuition the Divine Masculine is birthed, which becomes Intellect, which carries the descriptors to organize the divinely impelled Emotion into form.

In the womb, Intuition impels some stem cells to become bone and others to become hair, based on the complete plan of your entire life and all the potential of your mathematics. The steps are tiny, the plan is huge, and the miraculous, mysterious force of Compassion knows it all.

Intuition is also informed of what's happening on the surface of the planet through communication with the Observer. Observer feeds the mathematics of the Lens directly into the Source, thereby making that information a part of the gravitational wave flowing to the Lens. Intuition uses the information it receives to inform the next most advantageous step for the good of the whole—the whole plan, the whole body, the whole creation of present moment. It is associated with the heart region of the body.

From here, Compassion creates the third and most chaotic human dimension: Intellect. Here is where color and light emerge, bringing audible, organized thoughts, language, analysis, and descriptions of the physical world. Here the voice of self can be heard by self, and action is taken using free will. The form of the chair is created by Intuition enjoining the Emotion of the chair with the Intellectual description of the chair, which is then projected outside of you as a chair. Intuition has knowledge of the whole; Intellect is literally out of the loop, the last to know. But that will be our next lesson.

Looking from this perspective, all of the drama and suffering of daily life in the Lens is reduced down to a very small part of a great, miraculous whole. Oh, yes, there is a very real physical world. People are coming and going. Hearts are breaking and love is blooming. Jobs are being created and taken away. Money is coming in, or not. So yes, there is a physical world in the Lens with physical details that need your attention.

When you shift your attention, however, from the drama of the Lens to the entire map—this intricate, intimate perspective of how human life is created first from within you—it becomes undeniable that you are a miracle, just sitting in a chair, no matter what details are taking place in the Lens.

When you live in this broader perspective every day and take control over your Domain, the dramas in the Lens change. You experience how much power you really have to make a difference in your own life and you are inspired every day to use it. When you learn to use free will to make decisions for the greatest good of all, your projections change. The entire amazing mechanism of your Domain that is giving you the opportunity to have a physical life becomes the miraculous basis of your reality.

These pictures we are painting have never been articulated quite like this before. We hope this historical dissertation will change the way humanity understands itself. It's not the beginning, but it is a beginning of an in-depth articulation of personal empowerment that science and spirit have been building for many lifetimes. Even we who have been working through Barbara for many years, building our own vocabulary, doing our own exploration, studying you and your physical bodies, are on the brink of a new tomorrow.

Thank you for your continued participation.

Third Human Dimension: Intellect

Come with us now on a journey into perhaps the most complicated Human Dimension of all: Intellect. Position the map in front of you. See the center of the planet and the gravitational wave running from the Source through the Lens and out into the heavens, spinning up and around and back down again through the North Pole. Feel yourself at the Observer, as well as the Source, and fully present in your body sitting in the Lens.

The third Human Dimension of Intellect arises as Emotion fuels Intuition to impel the creation of form. As particles of Emotion hum, "I AM the chair," Intuition impels, "Go become the chair." Intellect then articulates a description of the chair based on its Source math that includes revolving the reference frame 180 degrees in preparation for the projection of the image of the chair into the outer world of the Lens. "I AM the chair" thus becomes "that is a chair," identifying the chair as something separate from and outside of self.

Intellect defines what is being created and perceived. Those definitions contain the measurements, location, judgments, qualifications, determinations, and instructions on how to interpret the creation from within the dictates of the culture of the spacetime in which the creation is being perceived. Waves of Intellect then proceed to mold themselves around these definitions and cement the tangible objects of life into their forms. Intellect is the only Human Dimension that possesses the power to analyze these spectrums of separation.

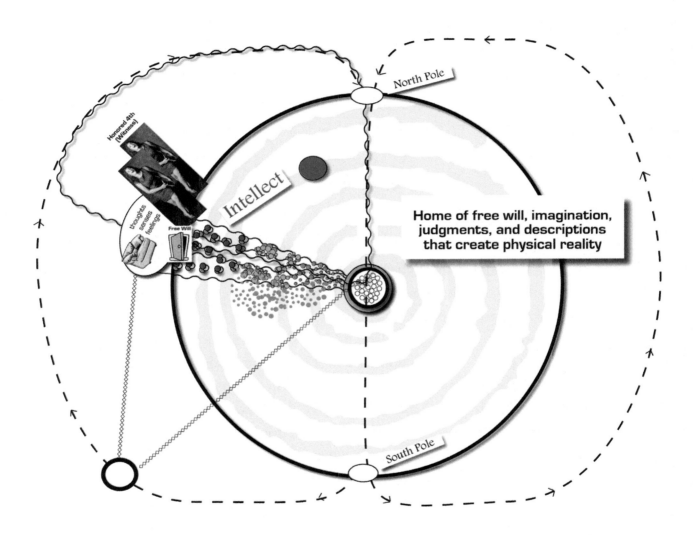

Third Human Dimension: Intellect

Act I: The Science of Compassion

When these three Human Dimensions lock together and pass into the Lens, physical reality is made manifest similar to how light flows through a prism and separates into a rainbow of color. Color is the by-product of light's interaction with the prism. The physical world is the by-product of your gravitational wave's interaction with the Human Intention via the human body in the Lens. All of the physical forms in the universe—the chair, the table, stars, cars, people, etc.—are energy that is being slowed down and separated into sound, light, and color, and projected through and perceived by your psyche in your body using thoughts, feelings, and senses.

Intellect is the home of free will: the power to have domination over self. As a human, you have the power to make conscious decisions. Free will is exercised in the Intellect and is what allows you to have the power to analyze the spectrums of separation and make decisions to take action. No other part of the whole possesses this power of free will and decision-making but Intellect.

Along with the manifestation of the body emerges *ego*: your psyche's non-physical self-identity in the form of a voice that defines the world to you as you move through it. Ego employs imagination and adds subjective and often judgmental definitions to the self-identity emerging into self-awareness.

You as a human have the power to give life and to take it away. You can use free will to decide to bring another human life onto the planet or to take your own life or that of someone else. This truth—the power of life and death—makes you a god of sorts.

Free will also gives you the power to decide to override the impelling of Intuition and make decisions that are not for the good of the whole. This ability to override Compassion is the primal root of all conflict that manifests in the Lens.

Intellect is associated with the head region of the body, where the mind is located.

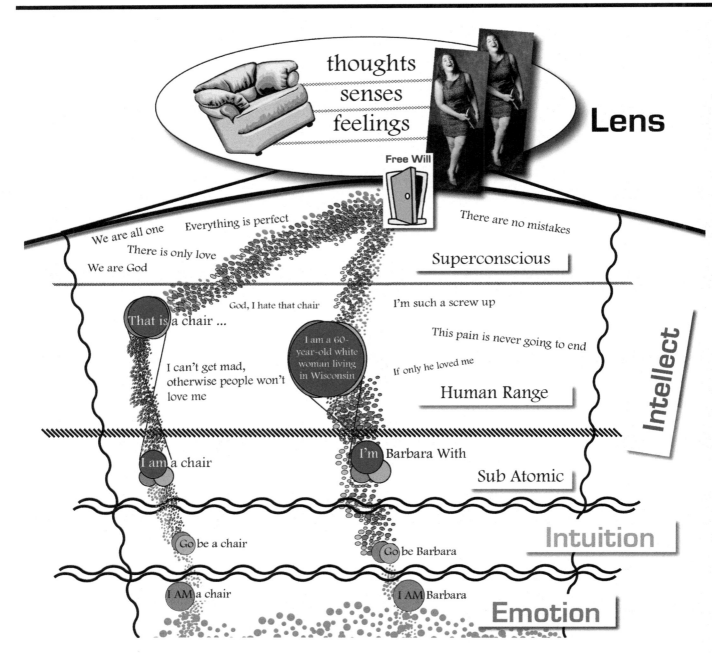

The Three Ranges of Intellect: Subatomic, Human, and Superconscious

Act I: The Science of Compassion

The Three Ranges of Intellect: Subatomic, Human, and Superconscious

Intellect contains three distinct frequency ranges within its energy field that serve three specific purposes:

Subatomic Range: All of the messages of the math of all physical objects that will be seen in the Lens begin to emerge in the subatomic range of Intellect. In the case of the chair, Emotion of the chair is vibrating with the mathematics, "I AM the chair. I AM the chair. I AM the chair." Intuition of the chair is impelling, "Go be the chair. Go be the chair. Go be the chair." As those two weld together and enter Intellect, a description emerges and form is born. The primal buzzing of the "I AM" of the chair can now be heard as a description of identity using language, "I am a green chair sitting in the corner of a room in Wisconsin in 2016."

The living mathematics in the Source are miraculously programmed to revolve the reference frame 180 degrees, instigating the emergence of a self-awareness that is able to witness the chair as separate. The definition of the chair now describes the chair as outside you via different latitude and longitude than where your body is. The subatomic Intellectual description, "I am a chair" becomes "That is a chair." Thus, you project the chair outside of you as something separate from you, complete with the entire description of the math of the chair, "That is a green chair sitting in the corner of a room in Wisconsin in 2016 ..." in the human range of perception.

Because of the cacophony taking place in the subatomic range of Intellect, a filter exists between it and the human range of perception. Imagine having to listen to the song of every object of the universe being sung deep inside your head! This filter protects the human mind from the insanity of hearing so many voices at once, which can blow neural networks apart. You don't plug your lamp into a power plant. Energy needs to be transformed downward from the Source until it can be safely accessed through a wall outlet or, in this case, the human mind.

Deeply embedded in the subatomic are also the stories of self-identify. The descriptions of who you are emerge from within to be projected into the Lens as and through a unique DNA body operating within a particular spacetime using thoughts, senses, and feelings to organize the creation of the physical world. Some stories of self-identity are objective and treated as fact. "I am a woman in the 18th century," "I am Native American," "I am a father," are examples of immutable inner truths of your identity, humming in the subatomic.

Other stories are subjective and are the result of the interaction taking place in the human range of Intellect between ego and the *voices of culture*.

Human Range: The subatomic voices of the I AM bubble up into the human range of perception, filtered through whatever spacetime culture the DNA body is experiencing and become what we call *voices of culture*. "I AM a chair" becomes "that is a chair" as the story of the "culture" of the chair and where it is located also emerges: "That is a green chair sitting in the corner of a room in Wisconsin in 2016 ..." These voices of culture are building blocks, helping pull together the forms that are emerging into perception by your body.

The ego is your personal voice of culture telling you the stories of your reality and your self-identity. This voice is shaped partly by the objective facts and partly by the subjective influence of the culture in which you live. In one community culture, thin women are revered; in another they are shunned. In one family culture, you are loved as a baby; in another you are abused. Those cultural messages affect the voices in your head as ego molds itself around the culture in which it is perceiving through your unique DNA, experiences, and reference frame.

Ego, operating in the Intellect, broadcasts the steady stream of self-talk you hear in your human range of perception that is constantly defining your world. Imagination gives ego the tools to add judgment to the description: "That is an ugly green chair sitting in the corner of a dingy room in that stupid state called Wisconsin in 2016 (god, am I old) ..." When you look at the chair, its reality is directly related to its subatomic mathematical definition bleeding up and revolving into your human range of perception to be projected outward. You may not always hear "that is an ugly green chair sitting in the corner of a dingy room …" going through your head as you look at the chair, but nonetheless you can access that definition in your human range of perception at any time.

With enough repetition, voices of culture in the human range of perception become embedded back in the subatomic. "That is a green chair," said to a child enough times becomes her language, embedded in her operating system. She learns to identify and define the world with her language. "That is a green chair." "I am a female." "These are my toes."

In the same fashion, the subjective definitions of ego can become embedded in the subatomic as well. If the human range of perception hears "You are not worthy of love" and is treated as such, the story "I am not worthy of love" can become a tape loop operating in the subatomic much like the song of the chair. "I am a chair, I am a chair, I am a chair" creates the experience of the chair. "I am not worthy of love, I am not worthy of love, I am not worthy of love" creates the experience of a life lived without love. You may not always hear "I am not worthy of love" going through your head, but nonetheless you can access that definition in your human range of perception at any time. How many times do you "beat yourself up" with self-talk that defines you as "not good enough?" This conflict taking place in the human range of perception has a profound effect on your relationship to what is manifesting in the Lens.

Superconscious Range: Here the most complete and truthful definitions are being generated into perceivable thoughts, ones that reflect the reality of the whole: that the entire mechanism is one unified field aligned to the M(iracle)-theory of creation being instigated into reality by Compassion. The superconscious range of perception provides access to the unified descriptions that contain the truth of the whole that are then used in aligning the three Human Dimensions to Compassion. This is also referred to as *Compassionate Intellect*.

The Conflicts of Intellect

Intellect vs. Intuition
Because of free will and imagination, Intellect can choose to ignore the impelling of Intuition. This happens when imagination analyzes the spectrum of any given moment in a subjective and degenerative way. Intuition can be impelling you to rest, and at the same time Intellect, filtered through the "I AM not worthy" tape loop, can be defining your reality as, "Nope, no time to rest. Have to work harder. Have to drive myself. Have to prove I am worthy." A tape loop of reasons why you cannot rest plays in your human range of perception (you can't let people down, you have to get this done, you can't fail, etc.). Anxiety builds in your body (tense muscles, stomach and headaches, fatigue, insomnia, etc.). In a knee-jerk fashion, you decide to keep working, thereby pushing your body and mental state into an even more weakened condition. Ignoring Intuition, Intellect makes decisions to take action only for the instant gratification of itself instead of for the good of all three Human Dimensions. This creates inner conflict and brings everything into further imbalance.

These decisions then become your reality. Ego defines your world through the filter of fear and lack, sometimes blaming someone or something else for your feelings or your depressed predicament. It punishes you for being such a failure, talking in circles, triggering more angst that fuels more knee-jerk decisions. Intuition can't even be heard. Emotion becomes abscessed. In this way Intellect can and does take you hostage, as the entire operating system is now controlled by the voices of culture instead of the voice of Compassion.

According to this perfect operating system, the conflicts between Intellect and Intuition will perfectly manifest in the physical world of the Lens. They will be present either in your body (illness, exhaustion, etc.), projected into the dramas of your culture (conflicts with friends, family, co-workers, wars on people and nature, etc.), in your quality of life (feeling unfulfilled and useless, living in constant depression and anxiety, etc.), or any combination thereof.

Intellect vs. Emotion

Emotion vibrates with the miraculous truth of who you are: "I AM a miracle." The immutable truth of M-theory is that the master planning and order behind the creation of human life inspires such awe and is rooted in such mystery that it can only be classified a miracle. The proof of this is you. The bottom line is you are a miracle of creation just sitting in a chair doing nothing. The pulsation of your gravitational wave is literally singing your body into being and filling it with your consciousness. It does not matter if you are thin, happy, poor, successful, straight, or depressed; it's a miracle that you even exist. So many energy systems mysteriously operate brilliantly together to create human life: neurological, biological, cosmological, physiological, psychological, quantalogical, etc. Who or whatever plans and implements this creation is truly a god. That is the truth.

Emotion, vibrating with this miraculous truth, then splits and welds itself to a particle of Intuition, which is impelling an action that will be best for the manifestation of the good of the whole mathematical scheme: "Go be that miracle!" When those two dimensions split again and reveal Intellect, the objective truth comes to life in the Lens as your body and the universe it is experiencing. You literally become the miracle: a woman sitting in a green chair in a room in Wisconsin in 2016 experiencing the universe.

The emerging Emotion, however, can attach to subjective and judgmental voices of culture that are embedded in the subatomic defining you as unworthy. The conflict is that you just *are* the miraculous act of creation, coming to life mysteriously in an all-abundant universe. You are not only worthy, you are priceless. There is no negating that. But the limited human ego is defining you as worthless and the world as a place of fear and lack.

Since decision-making takes place in Intellect, ego errs on the side of its own immediate gratification at the expense of the whole. These decisions then manifest as dramas in the Lens as you act out this worthlessness, fear, and lack. When Intellect denies this magical, immutable I AM and defines you as not worthy of your miraculous birthright, you become conflicted in ways that then influence your decision-making process away from the good of all. Imagine the different decisions you would make if you defined yourself as not just worthy of making miracles, but understood the extent of the power you possess and accepted responsibility that you impact the creation of your life with every decision you make.

Separation vs. Unity
Separation is a requirement to experience the physical world in the Lens. Consistently, Intellect empirically defines the physical reality as separate in order to experience it. Without that focus and those definitions, there is no physical reality. That is, after all, Intellect's job: to provide the definitions of the forms in the physical world.

But separation is a lie insomuch as it is not the whole truth. Despite the appearance of separation in the Lens, if you examine the entire system, you see one singularity with multiple reference frames. You are simultaneously separate from the objects and players in the Lens and unified with them at the Source.

Like a cataract on a lens that changes its opacity and leads to blindness, so, too, the voices of culture contain definitions that can make your human mind blind to the whole. Authority then ceases to be obtained from Compassion, originating in the Source, fueled by Emotion, and impelled by Intuition. Instead, authority is passed to a limited, subordinate system: Intellect, ego, and the voices of culture. The limits of your human perception then cause you to draw erroneous conclusions that cloud your thinking and influence the decisions that you make.

Since everything originates within you, the root of your part of any conflict begins within you. Even though the source of your conflict might appear in the Lens to be separate from you, it is originating within you. You are all the same organism, in the macrocosm and the microcosm of the same planet. Taking action to hurt someone else as a way to protect yourself ultimately only hurts you. You become the perpetrator of the very thing you think you are protecting yourself from. This is the real meaning of "an eye for an eye": not that if your eye is taken, you have the right to take another's eye. It means if you take an eye, that action adds itself to the math of your gravitational wave, only to come back around again. That will allow for an eye to be taken from you. This "us vs. them" mentality will destroy human life as well as the planet Earth if it is given free rein to escalate.

This is the power you have been ignorant of, ignoring and/or abusing. It's as if you're carrying a backpack that you aren't aware is filled with dynamite. Without knowing what could happen, you carelessly throw that backpack in your locker and dump it by the back door when you get home. If you knew the explosion you could cause because of your actions, you might act differently.

We don't say this because we think you're bad or wrong. We want you to understand the profundity of your power and to inspire you to use it in new ways. The good news is, the voice of your ego is a voice you have dominion over, whether or not you have ever exercised control over it. Learning to do just that is the work of Conflict REVOLUTION®.

The resolution of these inner conflicts is found in developing a present-moment relationship with all three of your Human Dimensions and exercising control over your own Domain. Free will gets trained to do the bidding of Intuition, and the deeply embedded, degenerative voices of culture are discovered and redefined in a regenerative fashion, thereby bringing all three Human Dimensions into alignment with Compassion (Cn^3). You then become a living embodiment of the truth of the whole while honoring and articulating the miracle of separation.

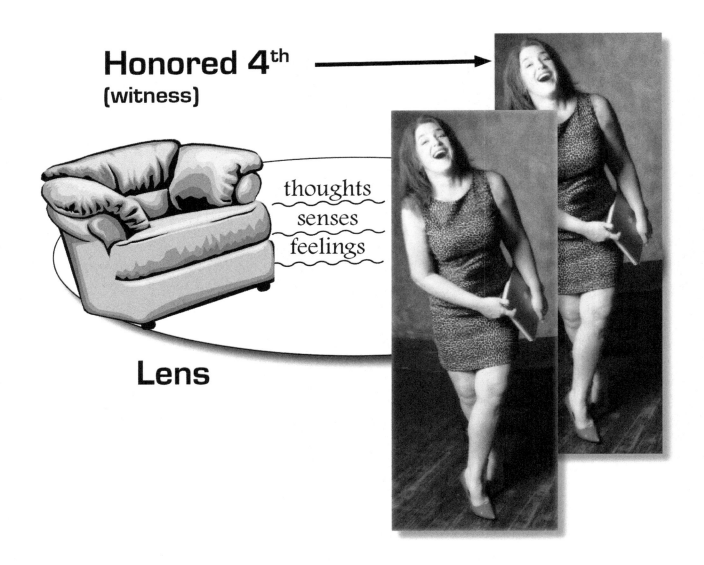

The Honored 4th: A Witness

The Honored 4th: A Witness

The Honored 4th is the reference frame of a witness, a self-observer that can perceive both the physical world of the Lens and the non-physical gravitational wave that precedes it. Here is where you become self-aware of what feelings you are experiencing, what Intuition is impelling you to do, what thoughts you are thinking, and what decisions you are ultimately making.

Like a witness called to testify before the court who can influence the outcome of a trial, the Honored 4th attests to what is *actually* taking place in the Lens (as opposed to how your ego may be defining it) and can thereby influence what decisions you will ultimately make. These new decisions have an effect on the outcome of what manifests in the Lens.

The Honored 4th is different than the Observer in that the Observer operates outside the free-will Domain. It is designed to transmit information about the conditions in the Lens back to the Source and experiences the body from the outside in. The Honored 4th operates within the free will of the Lens and is an observer of the body from the inside out.

Aligning to Compassion (Cn^3)

To align to Compassion, you watch yourself through the reference frame of the Honored 4th to examine intricately and intimately the condition of your three Human Dimensions. You bear witness to the decisions that you yourself are making in the Lens that are shaping your reality. You learn to understand the impact your actions have on the arena of your conflict, and how to employ simple, alternative, baby steps of change in order to gain control over your own decision-making process.

To align to Compassion, you process Emotion using your body and your breath instead of analyzing it using your Intellect. You learn exactly what Intuition sounds like and how to quiet the mind to hear it. You gain control over the wild horses of your thoughts and re-train Intellect to make decisions guided by Intuition.

These tiny steps have a huge impact on installing a new operating system. With enough repetition, this new operating system takes over and becomes as knee-jerk as the old. Except that in this case, decisions are now made for the good of the entire system, including what is best for you. In this way, you create the ultimate system of regenerative self-care that can't help but influence the world around you.

Evil

Just as we gave Compassion a new, less-subjective definition, let's now give *evil* the same. For our purposes, evil is merely *decisions made for the good of the few at the expense of the whole*. Every human has the potential to act in this "evil" way.

Do not be fooled: you cannot protect yourself from evil by destroying those things outside you that you define as "threats" and the source of your evil. Destroying those who you deem are evil is like breaking the mirror because you don't like what you see in it. First and foremost, you must find the evil within *you* and change it from there.

With this new, nonjudgmental definition of evil, you must find within yourself the times you make decisions for the good of the few (only your ego) at the expense of the whole (your body, your mental condition, other people, etc.). You must take control of your own Domain and then intend to bring unity, first and foremost, to your inner world.

Act I: The Science of Compassion

This kind of self-scrutiny is needed in order to evolve as a species. This crux of action is where the change must take place to transform your world at its roots. Aligning with Compassion guarantees that any decision you make will be made for the good of the entire situation—not just to appease your ego—and will positively impact the outcome of every circumstance and situation you find yourself in.

Your mission is to become aligned with true north of Compassion: to experience the reality of the Source and the Observer as equally as you experience the reality of the physical world being projected through your body inside the Lens. Compassion—the self-awareness of the parts that the parts all originate from the same singularity—always impels decisions based on the good of the whole.

When your consciousness of pure energy can perceive itself for what it is while cemented in the matrix of a physical life, you will truly make this evolutionary step.

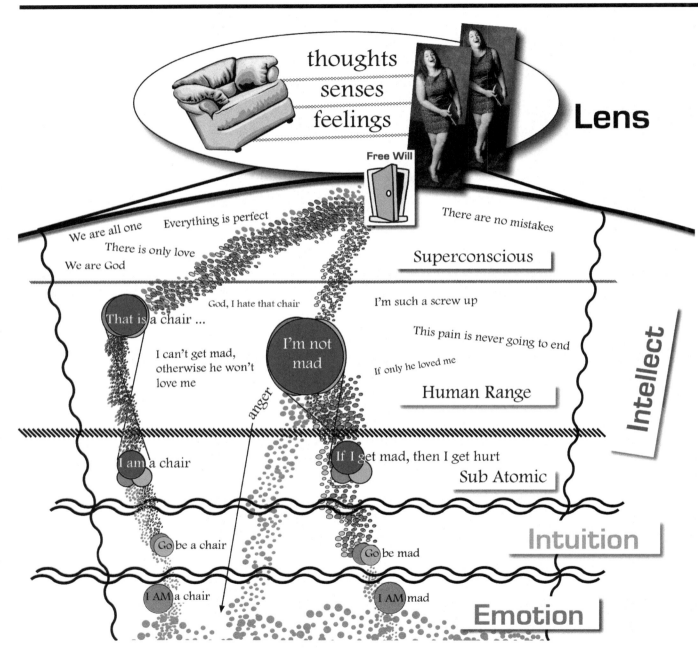

An Abscess of Emotion, View 1

Abscesses

To be aligned to Compassion, all Emotion needs to be honored and flow through the physical body to continue the spinning journey back to the Source. When Emotion does not flow through the body, it becomes trapped in an abnormal cycle of backward flow called the *Abscess*.

The very basis of experiencing physical reality is reliant upon Emotion teaming up with Intuition and Intellect to be projected into the Lens as your body perceiving the surrounding physical reality. The green chair in the corner begins within you as the Emotion of the chair—"I AM a chair"—that weds to the Intuition of the chair—"Go be a chair"—that Intellect consequently defines and revolves to become "that is a chair" and voilà—your physical body perceives the green chair in the corner of the room in Wisconsin in 2016.

Emotion fuels the creation of the physical world. One end of the Emotional spectrum has no more or less value than the other. Imagine saying that red is more important than blue in the color wheel. You need all three primary colors to make the infinite spectrum of color that is used to perceive the physical world. The same is true of Emotion: to live aligned to Compassion you need to find a way to move the entire spectrum of Emotion through your physical body and prevent abscessing.

In today's culture, however, you are told to stay away from "bad" feelings like anger, sadness, and depression. "Good" feelings like happiness and joy are valued. In reality, culture today generally does not discern between Emotion and Intellect.

When you separate the *feeling* of rage from the *thoughts* of how to act on the rage, you can access your power to change. You can begin to process the feeling of rage in one way and the thoughts about what to do about the rage in another.

The truth is, there is no "bad" Emotion. Anger is no better or worse than joy: it's what you decide to do with the anger that will determine its influence on the environment and your life.

A misdiagnosis of conflicts of the psyche is falsely believing that anger must equal aggressive action against another or self. That leads to falsely believing that you cannot allow yourself to feel the anger, lest you do something hurtful or destructive. That definition serves to block the anger from flowing through your body and deflects it back into the Abscess.

Say as a child, whenever you expressed anger, you got smacked. This embedded a math of sorts into your subatomic—"If P, then Q"—that becomes, "If I get mad, then I get hurt." The conflict arises of having to choose between being true to self or self-survival. This, in essence, is about not valuing the feeling of anger: it's not worth feeling anger due to the imminent danger it puts you in. This revolves into "I am not mad (therefore I cannot get hurt)." As this imperceptible, repetitive subatomic message weds itself to a less "acceptable" feeling like anger, it is driven into reality by Intuition ("Go be angry"). "If I get mad, then I will get hurt" gets revolved and projected into "I am not mad." The resulting condition defines you as someone who is not mad, even though in reality, anger is moving up the gravitational wave, only to be blocked by the overriding definition of being someone who is not mad.

This definition—"I am not mad"—works like a cataract, deflecting anger from getting through into the physical body. However, that Emotion does not dissipate. Anger that is not able to continue its journey through the body and onward into the heavens gets trapped on the gravitational wave in an unnatural backflow—the Abscess. This trapped mass of unexperienced Emotion is then perfectly made manifest in the Lens as conflicted form: physical illness, such as an actual abscess in the physical body; recurring conflicts within relationships; or any number of other blockages to health, happiness, and regeneration. All the while, your ego is out of touch with the anger and genuinely appears to not feel a thing.

Imagine the color wheel without red. It would be an entirely different world. Humans must learn to experience all colors of the Emotional color wheel in regenerative ways. In order to

Act I: The Science of Compassion

allow all Emotion through, your job is to redirect Emotion through the body by marrying it to messages of new, Compassionate Intellect, and to discover the degenerative, subatomic voices of culture and rewrite their stories.

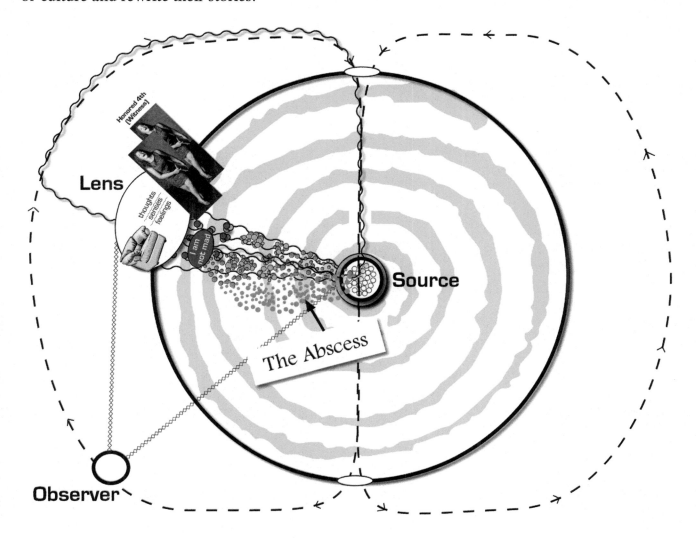

An Abscess of Emotion, View 2

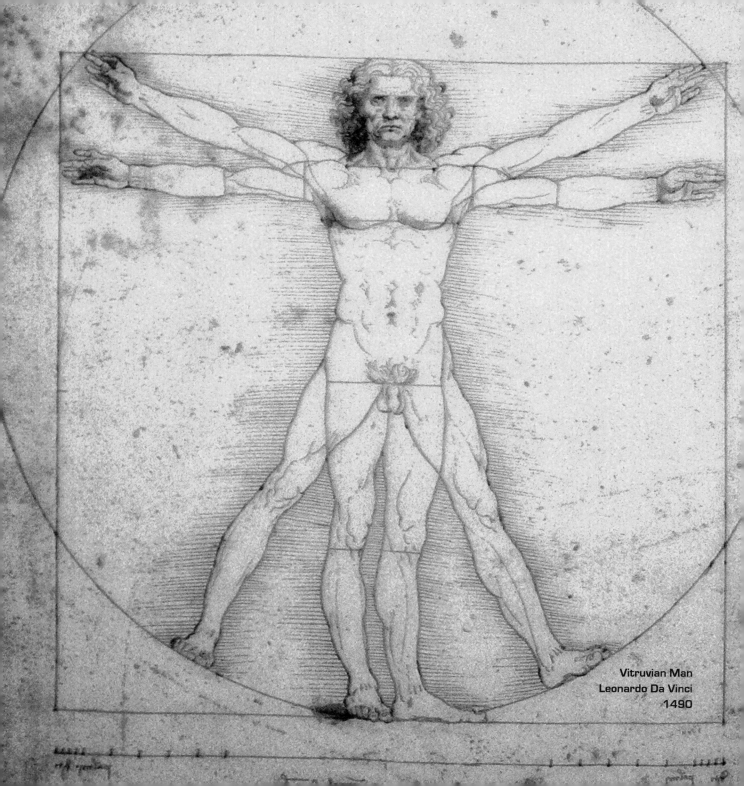

The Human Body

Again we begin with a mystery: who or what decided that you would have ten fingers and ten toes, arms, legs, a heart, a brain? Every part of your body begins as the same kind of cell. So who or what inspires one stem cell to organize and divide into skin, another into bones, yet another into hair or blood? It is Compassion and you are a major part of the implementing design.

Like snowflakes, there will quite literally never be another you. The math of your Human Intention determines the unique angle at which your consciousness will be projected into the Lens. By determining what DNA body you will be creating in which spacetime and at what latitude and longitude, you also determine who your parents will be and therefore what physical features, such hair color, body shape, skin tone, and genetic predisposition you will possess. Moreover, this inner gravity of Compassion determines a general structure for how your life will unfold.

Pre-Conception Planning Meetings

Before you are conceived in your mother's womb, the consciousness of all who will be participating in the creation and implementation of your life gather and, like corporate negotiators in a boardroom, make contracts and chart destinies. In these negotiations, it is not that you, for example, ask the soul of your future father to beat you or request that he abandon you. The math enclosed in your compilation will determine the conditions under which you will interact together. So instead of specific decisions and actions being plotted, such as beating or abandonment, a destiny is formed—a matrix that becomes the guide to determine the growth and development of each of your lives on Earth. Within this matrix, the living math in the Source keeps the plan on track.

The math of your gravity dictates much of the plan, but free will allows you input into how that plan unfolds once you have been born. There will be many opportunities to make free-will choices along the path that is laid out for you. So instead of programming that your father beat you, there is an implicit agreement between you and your father to live within certain physical conditions (poverty, disability, mental illness, wealth, etc.) that are designed to allow consciousness an opportunity to evolve. Those conditions may provide your father the impetus to physically abuse you, but they also provide him with the opportunity to rise above and understand those conditions from a more evolved perspective, thereby influencing him to make different decisions.

The same is true of your life. Your father might react to his physical conditions by beating you, but you, too, have been given the power to choose. You are given the opportunity to make any number of choices about your actions and reactions to the interactions.

Life on Earth is a beautiful dance of great cooperation; it becomes the sum and consequence of all these decisions that you make. They are recorded on your gravitational wave, which then changes the math back at the Source. At any time in your life, you are able to change your perspective and thus your actions and reactions to anything you face, thus changing the math in the Source. This element of free will within destiny is the key that will eventually lead you to understand the power you possess to influence how matter manifests on Earth.

Birth

Emerging from the nothingness of the watery womb comes your body, your mysterious, miraculous body, the physical expression of your eternal consciousness. You help to construct your fetus, your consciousness darting in and out of your mother's womb like a bird building a nest. But your energy is only one of many creating the fetus.

The consciousness of your father and mother each create a sperm and an egg. Those then create their own "Big Bang," and Compassion incites matter to cluster to become your cells, which form your bones, muscle, flesh, and eventually all of the intricate, intimate details of your physical body and psychological mind. Intuition determines which frequencies will vibrate which stem cells into becoming which form, drawing from the special blueprint of you based on the plans from the boardroom negotiations.

Nestled in your mother's womb, the genesis begins. Your growing fetus is completely dependent upon your mother's ability to nurture your fertile intention. When you finally make the journey down the birth canal and take your first breath, your consciousness becomes cemented into your physical body. Since you have literally been growing underwater in the womb, you have not yet experienced separation.

With your first breath a boundary of skin appears where your body stops and the air begins. This boundary defines what is contained within you and what appears to be outside you. Human Intention also separates and reveals Emotion, Intuition, and Intellect. Human Affiliation arises as your psyche splits into Compassion, Vision, and Honor & Articulation. Human Perceptors of thoughts, senses, and feelings are enabled.

As a newborn baby, you experience life primarily from Emotion and Intuition. Emotion as the primal I AM of the entire organism inspires Intuition to impel the next most advantageous step: to go be you, to eat, sleep, cry, smile. There are no subjective stories or voices of culture swimming around in your head yet. There are no judgments telling you something is good or bad, right or wrong. Objects are defined by the messages in the subatomic. So as a baby you hear the song of the chair radiating from within you, and like magic, outside you a chair appears. You become caught up in the present-moment experience of the creation of the chair.

When the bonded Emotion and Intuition enter the subatomic Intellect, the I AM of the electromagnetic Domains of everything making up your physical world—the math of the chair, floor, spoon, your mother, even your own body—vibrate into form at this quantum level. Your human range, which contains the revolved definitions, is focusing on the outer world and projecting these emerging forms into it.

Your psyche is also being influenced by the voices of the culture of the spacetime you are in. You learn the language and customs of your locale and time period, which embed themselves into the subatomic range of your operating system and become second nature. Definitions of who you are arise based on your predetermined math and its relationship to the voices of the culture in which you are living.

At first it appears you have no control over things that happen: when your parents feed you, when you get put down for a nap, or who holds you and moves you from place to place. You quickly learn your first "If P then Q"—the first equation of the math of your universe: if you cry, then … what? Someone comes and feeds you? No one comes to feed you? This is the genesis of the matrix of your plan. These are the first voices of culture becoming embedded in your subatomic. You learn that in your house, if you cry, someone comes and picks you up. Or you learn if you cry, someone hits you. These early embedded equations invisibly influence the direction of your life and your free-will choices.

If the culture you lived in as a baby taught you that if you cried or got angry, you got hit, over time the "If P then Q" of "If I get mad, then I get hurt" becomes embedded in your subatomic. Revolved 180 degrees into the human range, the message becomes "I am not mad" as a way to protect the mechanism from harm.

As your body continues to grow, so does your awareness of your own power. Crawling, talking, learning the language of the faces in front of you, you gain control of more and more of your physical world. Mama. Cookie. Dad. No. Bad. Good. Right. Wrong. You mimic the repetitive sounds you hear in culture that identify and define your world until you know the language, thus cementing the reality of separation a step further.

As Intellect begins to develop, you start to think for yourself and separate from the culture of your mother. You merge your power of deduction with your free-will ability to take an action: if I crawl across the room, then I can have my toy without crying for mama. At some point, you become aware that the decisions you make are shaping your life.

Act I: The Science of Compassion

As the ego develops further and embeds more and more definitions from the spacetime culture into the subatomic, a quandary is created. The subatomic range is still defining you as a miraculous, present-moment creation, a wave of consciousness pulsating out of a primordial soup with the song of the universe, all parts in harmony with themselves and each other, singing itself to life through your body as one miraculous unified being.

And yet when your ego contemplates the ideas of being a miracle, of everything being one, "you" are clearly not "me." You can't run your hand through tables or walk through walls, no matter how hard you might try. And you certainly don't feel like a miracle just sitting in a chair most of the time. To the ego, the limited empirical experience of this separate, hard reality of matter overshadows the truth of the unity of the Source. This constant conflict becomes a part of the operating system of your body.

Make no mistake: life on Earth is a beautiful dance of great cooperation. The master planning and order behind the creation of the life of a human body can only be classified a miracle. At any time in the process of being alive, you have the power to change your perspective and thus your actions and reactions to anything you face, which in turn affects the details of the Lens in which you are having the miraculous human experience.

So how do you redefine yourself as a miracle of creation just sitting in a chair doing nothing? How can you break free from the subatomic messages that are creating constant conflict in the very root of your human experience?

This is the work of Conflict REVOLUTION®. When you redefine your identity by re-channeling Emotion that has been trapped in the Abscess and unite it with superconscious thoughts, anything is possible.

Flat Einstein
Palace Hotel, Madrid Spain
December 2014
Photo: Barbara With

Einstein in Madrid

After his successful visit to Paris in 1922, Einstein journeyed to Spain to do a similar such peace mission in Madrid, Barcelona, and Zaragoza. Here he lectured, attended receptions under great public and media scrutiny, and made several cultural visits, including the Museum del Prado and El Escorial.

Einstein's new theories were inspiring the world—everyone from philosophers to politicians was intrigued with his new view, so simple and elegant: $E=MC^2$. Because of this, he was the equivalent of a rock star of the times.

While in Madrid he spoke at the Palace Hotel, meeting with the rector of the Central University and the faculty of the School of Sciences. A large banquet was held after the day of presentations, where Einstein was honored with a tribute to his genius and revolutionary leadership.

Act II

Conflict REVOLUTION® Learning Project

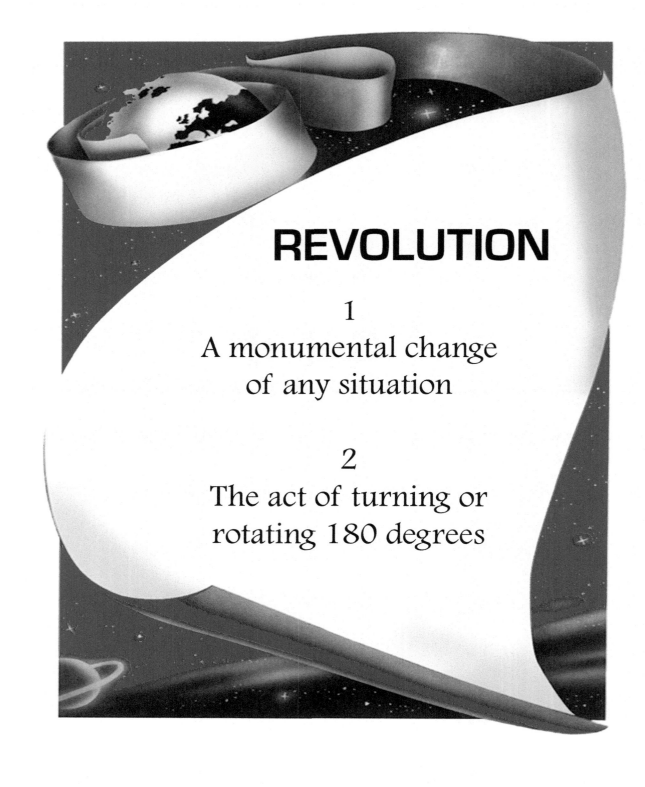

REVOLUTION

1
A monumental change of any situation

2
The act of turning or rotating 180 degrees

Creating a Unified Field

Conflict REVOLUTION® is a process of self-scrutiny that entails examining your Domain to find the imbalances within that are contributing to the conflict you are experiencing in the Lens. The reference frame of the Honored 4th allows you to monitor the condition of your Human Dimensions. From that perspective, you resolve the conflict there by quieting the Intellect, replacing the lies and voices of culture with nonjudgmental voices of the superconscious—the voices of truth. Then feeling and breathing present-moment Emotion, you move it through your body, leaving room for Intuition to influence free will to do what's best for the good of the whole of any given moment. This creates a unified field and what we refer to as *consciousness to the third power* or Cn^3. This powerful place of alignment to Compassion naturally influences your relationship to what is manifesting in the Lens.

Intend Change

Conflict REVOLUTION® is a journey of discovery. Its practice can be the hardest, yet most rewarding, work you will ever do. After a while, the new decisions become second nature. It's as if you've installed a new operating system that corrects the imbalances as you go by making different decisions. You become Intuitively guided and choose to take the baby steps needed to organically bring into reality your intention to do the greatest good—to "go be the miracle."

To learn this revolutionary process, we ask you to bring a live conflict to the table. You will use the details of this conflict to begin to separate Intellect, Intuition, and Emotion and create an Action Plan to resolve the conflict at this root level. Pick a conflict, if possible, that has not been resolved fully, if at all.

Don't be too concerned with picking the "right" conflict. No matter which drama you are drawn to examine, all roads lead to the root of the conflict, which generally has little to do with the story you are telling about it. Those details are meant to lead you back to your Domain, which is your responsibility and what you have dominion over.

The Benefits

Alignment to Compassion. You redefine "Compassion" and become the very change you wish to see in the world around you by choosing to exercise Cn^3 in your daily life.

Taking care of yourself. You are no longer dependent on others to take care of your conflicts. You can resolve them without anyone else's assistance or participation. This is self-love.

Increased energy and inner peace. Once you stop expending energy on emotionally attaching to life's degenerative definitions, you have more energy to spend on manifesting your needs and giving back to the world.

Fewer internal conflicts, therefore fewer external ones. As per the reflection of your inner world, so will your outer world become more peaceful. People will either shift as you do or find someone else to engage in their dramas.

More creativity. Once you start exercising your creativity in learning a new way to resolve conflicts, that creativity is available for other pursuits as well.

Taking action to influence peaceful outcomes. You will no longer sit on the sidelines feeling like life is in control of you. You consciously and intentionally take actions to manifest peaceful outcomes to everyday conflicts.

Becoming a leader; inspiring others. By taking control of your own Domain, you become an inspiration to others and show them how it can be done.

Happiness. More everyday moments are spent being happy instead of obsessed and conflicted.

Contributing to world peace. As you create peace within, you manifest peace in your local universe, doing your part to contribute to world peace.

Mission Statement

World peace, one person at a time, starting with self.

Goal

To learn a revolutionary new process to resolve conflict and use that process in everyday life.

Objectives

To identify, understand, and experience the components of your Domain—their form, function, and relationship to each other as seen through the reference frame of your Honored 4th.

To learn to use Conflict REVOLUTION® using the details of a real-life conflict.

To develop a self-awareness of your Domain and change your own decisions in order to align to Compassion (Cn^3).

To choose to use Conflict REVOLUTION® in everyday life.

Ground Rules

1. *Your* Domain is your responsibility
You are here to focus on your Domain, not that of your neighbor, your father, or anyone else. *Your* Domain is your responsibility.

2. Your Domain is *your* responsibility
No one else can take care of your Domain for you. It is your responsibility to take care of it. Your Domain is *your* responsibility.

3. See 1 and 2.

Values

Passion
You not only have every right to feel all your feelings, it is your personal responsibility to feel all Emotion and make sure your feelings are flowing through you.

Creativity
Creativity is your birthright. We highly value using imagination to find creative solutions.

Nonjudgment
This is a science project. Objectivity is crucial.

Humor
If you can't laugh at yourself, we will do it for you.

Conflict Quiz

Please answer the following questions as honestly as you can:

1. Which of the following best describes your relationships with your friends?
    a) I have no friends.
    b) I have a few close friendships that are easygoing, few conflicts.
    c) There seems to be a repetitive theme of conflict that runs through most of my friendships.
    d) My friends and I have conflicts but we somehow manage to resolve them.
    e) I generally keep people from getting too close to have conflict in the first place.

2. How would you describe your relationship with your co-workers?
    a) Several people bug me big-time.
    b) I get along well with most people, am easy-going, and have few conflicts.
    c) There are a few specific people I don't get along with whom I just steer clear of.
    d) I appreciate everyone I work with and have no conflicts with anyone.
    e) Everyone's an idiot.

3. How happy are you with the physical structure of your house and home?
    a) Better than sleeping under a bridge but not much.
    b) It's okay, will do for now.
    c) Somewhat but would like to make improvements.
    d) Have to find another place as soon as possible because living here is driving me crazy.
    e) Very.

4. How would you describe spirituality in your life?
    a) I am very strong in my belief systems and use them in my everyday life.
    b) Spirituality completely eludes me.
    c) I have doubts if there is anything spiritual, but admit there is much I don't know.
    d) I know there is something out there, but I'm not sure what or how to get it.
    e) I believe in the power of a spiritual life and am seeking ways to incorporate it into daily life.

Act II: Conflict REVOLUTION

5. What would you say is the condition of your health?
   a) I have several problems that I'm trying to cope with which decrease my energy regularly.
   b) I'm generally healthy, but suffer from a few recurring problems that are manageable.
   c) My body is falling apart, and I don't know what to do about it, nothing seems to help.
   d) I am healthy, physically fit and active, and live a high-energy life.
   e) I'm such a mess, I wish I were dead.

6. When it comes to money:
   a) I'm broke all the time, no matter what I do.
   b) I make little money and always seem to be scraping to make ends meet.
   c) I have a steady income but still seem to live from paycheck to paycheck.
   d) I have a steady income and a strong savings but am still often concerned about finances.
   e) I am happy and secure and rarely worry about money.

7. When it comes to having fun:
   a) I rarely have fun, but when I do, I realize how little fun I have in life.
   b) I find that life is fun and I'm always looking for new and stimulating things to do in my free time.
   c) I like to have fun, but need to find more creative ways to have it.
   d) I don't have much fun but I experience times that are mildly amusing.
   e) Life sucks.

8. Which of the following best describes your relationship to your family of origin?
   a) I don't like them, and if they weren't my family I wouldn't have much to do with them.
   b) They are my source of strength and support, and I regularly spend quality time with them.
   c) I try to keep my distance from them because we don't always get along.
   d) I wish I was an orphan.
   e) I appreciate them in my life and spend occasional time with them.

9. Which of the following best describes your relationship with your significant other?
   a) I love my significant other but we need to work through some things that continue to come between us.
   b) I just left my partner and am happy to have him/her out of my life.
   c) My partner and I are having problems and need some serious intervention.
   d) I love my significant other and feel blessed to have him/her in my life.
   e) I'm presently in the process of trying to leave my partner.
   f) I have no significant other.

10. Which of the following best describes how you deal with conflict in your life?
    a) What conflict?
    b) I don't like to deal with it but when it happens I try my best to take responsibility for my part.
    c) I find dealing with conflict exhilarating and a good opportunity for learning.
    d) I hate to deal with conflict and will avoid having to face it at all costs.
    e) I don't like to deal with it, so when it happens I try to avoid facing it.

Act II: Conflict REVOLUTION

Score the test as follows:

1. a) 5
   b) 1
   c) 3
   d) 2
   e) 4

2. a) 4
   b) 2
   c) 3
   d) 1
   e) 5

3. a) 4
   b) 3
   c) 2
   d) 5
   e) 1

4. a) 1
   b) 5
   c) 4
   d) 3
   e) 2

5. a) 3
   b) 2
   c) 4
   d) 1
   e) 5

6. a) 5
   b) 4
   c) 3
   d) 2
   e) 1

7. a) 4
   b) 1
   c) 2
   d) 3
   e) 5

8. a) 4
   b) 1
   c) 3
   d) 5
   e) 2

9. a) 2
   b) 5
   c) 3
   d) 1
   e) 4
   f) 0

10. a) 5
    b) 2
    c) 1
    d) 4
    e) 3

Conflict Quotient

Score:

10–18 points: Congratulations! You have a natural inclination for responsibility and resolving conflicts. With good guidance and learning from your mistakes, you have created a life that is fairly peaceful. Learning more about how to handle conflicts will enhance the strengths you already have.

19–36 points: While you are successfully dealing with some conflicts, you are still in need of more understanding about where you have not been showing up fully in your life. Learning more about how to handle conflicts will allow you to participate more actively in your life.

37 or more points: Your life is chaotic and often on the verge of crisis, depression, anxiety attacks, and general feelings of hopelessness. Immediate attention to how to handle conflicts will allow you greater control over the chaos and your Domain, where you can start to live the life you long to live.

Notes

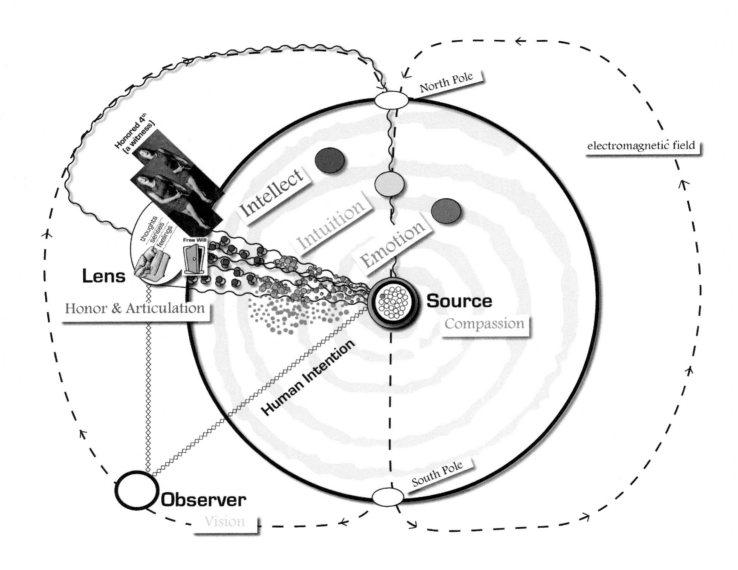

Map of Human Consciousness: Your Domain

Your Domain

YOUR DOMAIN is a 13-dimensional operating system originating in the Source and aligned to the electromagnetic energy of Earth that is responsible for your ability to experience the physical world. Dimensions include:

- Human Intention: Source, Observer, and Lens
- Human Dimensions: Emotion, Intuition, and Intellect
- Human Affiliation: Compassion, Vision, and Honor & Articulation
- Human Perceivers: feelings, senses, and thoughts
- Honored 4th, the Witness

YOUR DOMAIN is much bigger than your conscious mind realizes. Your Domain is like being your own planet, filled with mountains, oceans, and vast expanses, but only being able to experience ten acres of it (the Lens).

YOUR DOMAIN is your territory and realm of control.

Conflict REVOLUTION® redefines who you are based on this bigger picture and teaches you to exercise control over the dimensions that are yours to have dominion over.

Compassion: A Scientific Definition (Page 17)

Four Fundamental Forces of the Universe

1. Electromagnetism: Igniting
2. Gravitation: Guiding
3. Strong Nuclear Force: Attracting/Repelling
4. Weak Nuclear Force: Transforming

Fifth Fundamental Force of the Universe

5. Compassion: Intelligence that uses the four Fundamental Forces to impel creation of the physical world one step at a time.

Act II: Conflict REVOLUTION

Compass: The Source

A device with two arms used for drawing circles, one arm anchored in the center, freeing the other to pivot and create the physical manifestation of the circle. Compassion impels energy to step out of the zero point and into all potential to create your compilation of consciousness—one point within an electromagnetic particle in which all of the mathematics of your life experience are contained.

An instrument that defines geographic direction using electromagnetic fields that provoke a needle to align with the electromagnetic field of the Earth. Compassion impels consciousness to align with the electromagnetic field of Earth in order for all emerging humans to be on the same map.

Compassion impels consciousness to possess a self-awareness on the parts of the whole that all parts originate from the same Source.

Human Intention (Page 25)

The Prism

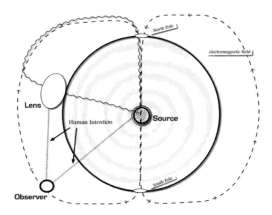

HUMAN INTENTION is a triangular energy structure consisting of three points—Source, Observer, and Lens—that operates like a prism, providing substance and separation to the light and sound particles on your gravitational wave to reveal the three Human Dimensions—Emotion, Intuition, and Intellect.

HUMAN INTENTION is consciousness' intention to create a body with thoughts, feelings, and senses to be alive in the Lens of the physical world.

HUMAN INTENTION manifests your body as per your math, including spacetime coordinates of where you are on the map and the DNA of the body that you are creating.

HUMAN INTENTION can analyze the spectra of the three Human Dimensions, allowing humans to be self-conscious. This is the "human" aspect of your experience, as opposed to the chair, tree, or star. Only human beings possess the power to analyze the spectra of self from their physical reference frame.

Act II: Conflict REVOLUTION

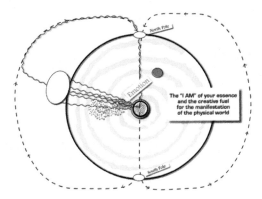

Emotion (Page 39)

The I AM essence and creative fuel of manifestation

EMOTION is the creative fuel of the physical world and the I AM essence of who you are. Emotion is associated with the the first point of Compassion and the Source. It is the primordial soup of the physical world begins to emerge and is located deep inside the Earth.

EMOTION flows through your physical body and shows up as feelings that, along with senses and thoughts, create your present-moment experience in the Lens.

EMOTION has no language or Intellect, no comparisons, descriptions, or judgments. There is only the pure experience of raw Emotion of who you are in any given present moment, translating into feelings: sadness, grief, joy, anger, happiness, frustration.

When EMOTION does not flow through the physical body, it becomes trapped in an abnormal cycle of backwash called the Abscess. Abscessed Emotion builds up and eventually becomes overwhelming, manifesting depression, physical problems, lack of energy, etc., and contributing to all the conflicts you are experiencing.

Intuition (Page 45)

Small declarative statement impelling the next most advantageous step for the good of the whole

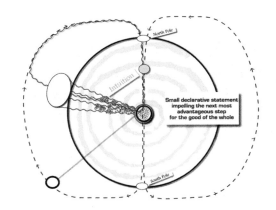

INTUITION is the voice of Compassion, sending an impelling wave onward to the Intellect that articulates the next most advantageous step for the good of the whole:

INTUITION can be a subatomic message ("Heart: beat!"), a physical feeling ("chicken skin," a gut feeling, etc.), an unscripted action (stopping at a green light, only to have a car run the red light that would have hit you), or a language message relayed to Intellect ("Call your mom," Take the job," "Turn left").

INTUITION determines this step based on the condition of the whole mechanism, which includes what is transpiring in the Lens in present moment, as being reported via the Observer to the Source.

INTUITION accesses information that is deeper than Intellect can comprehend.

Act II: Conflict REVOLUTION

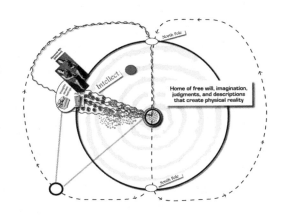

Intellect (Page 49)

Home of free will, imagination, judgments, descriptions, and stories that define physical reality

INTELLECT contains all the definitions of the emerging physical world.

INTELLECT contains ego, which uses imagination and language in a constant stream of self-talk to make judgments and comparative analyses of what is being created in the physical world.

INTELLECT originates the voices of culture. Ego tells the stories of who you think you are. Some parts of the story are *objective* and are treated as fact: I am a woman, black, a mother, homeless. Other parts are *subjective* and treated as opinions or judgments: I am ugly, competent, attractive, worthless.

INTELLECT is where conscious decisions are made by exercising free will.

INTELLECT has the power to make decisions that override the impelling of Intuition.

Degenerative Voices of Culture:

Mind Reader fabricates a story about what someone else is thinking and what their actions mean. Mind Reader is overly concerned with what she thinks other people think of her, and it's usually bad.

Critic is full of negative, judgmental stories that are sometimes meant to punish. "I'm no good," "You will never amount to anything." "I can't change." Critic's self-esteem is usually based on these statements, which are focused on human imperfections instead of the miraculous system of creation. These critical judgments prevent unification, create severe limitations, and, along with blame and resentment, get projected outward onto others.

Comparer constantly compares herself to another thing or person and judges herself as either better-than or worse-than. This causes a constant emotional roller coaster, switching back and forth, first feeling superior to others, then inferior. Comparer's self-esteem is dependent on her opinions of others.

Perfectionist has "All-or-nothing" thinking that says she is either entirely perfect or entirely worthless. There is no middle ground. If she wants to maintain her perfection, there is no room for mistakes. But if she can't make mistakes she can't grow, so Perfectionist is trapped in a continual sense of failure. Even when goals are set, they are often unrealistic and connected to an "if/then" scenario: "If I can become a really nice person all the time, then everyone will like me and no one will ever get mad at me."

Crisis Maker / Drama Queen makes every challenge an insurmountable crisis, bound to come to no good. Drama Queen creates crisis because it's what she knows and how she was raised. It's a comfort zone. Crisis Maker burns energy putting out fires she lit in the first place.

Honored 4th (Page 61)

The objective witness

HONORED 4th is an objective witness of the three Human Dimensions that operate in your body within your Domain.

HONORED 4th operates within the free will of the Lens and can influence decision-making.

HONORED 4th, like the director of a play rather than the actor on the stage, can step back and analyze what's taking place within the Domain from a place of detachment and nonjudgment.

HONORED 4th distinguishes between the stories in your head of who you think you are and who you really are based on observation of your actions.

Like a witness called to testify before the court who can influence the outcome of a trial, HONORED 4th attests to what is actually taking place in your Domain and possesses the power to influence free-will decisions.

The Road Map

Using Conflict REVOLUTION®, we will build a Road Map to help you observe the decisions that you yourself are making that are contributing to your conflict. This Road Map then becomes your Action Plan, with concrete steps to correct the imbalances in your Domain.

By following your Road Map, you will scrutinize the relationship between your Intellect, Emotion, and Intuition. By adopting the perspective of a witness, you will in effect catch yourself in the act of perpetrating the very conditions of your Intellectual stories. You will find where you are making decisions for the good of the few (ego) at the expense of the many.

Catching yourself in the act and choosing to make a different decision creates lasting transformation. This is what it means to create a new operating system.

To help you learn the actual process of revolving your conflict, here are some other terms to assist you with your work.

Projection, Triggers, and Detachment

Projection: Defining the source of your thoughts, senses, or feelings as someone or something outside you and projecting subjective, judgmental, degenerative thoughts onto yourself or others. You cannot escape projection. Physical reality is created in part using projection. Emotion, Intuition, and Intellect are projected from within your body outward using thoughts, feelings, and senses to create the experience of a world separate from you.

While you cannot completely divorce yourself from projecting, you can learn when your ego is projecting subjective and erroneous stories onto something or someone outside of you and/or when you are projecting degenerative and judgmental stories onto yourself.

Act II: Conflict REVOLUTION

Trigger: A person or event that appears to cause an emotional reaction (triggering event) and the nature of the emotion itself: anger, hurt, offense, fear, etc. (triggering emotion). Becoming triggered—having an abrupt emotional reaction to a person or event—is the red flag signaling the need to adopt the reference frame of Honored 4th and prepare for a Revolution.

In our step-by-step process of creation. (See page 90.) The triggering emotion emerging from the Source precedes the triggering event emerging in the Lens. Since Emotion is the very fuel that is creating the event, the anger, fear, and/or resentment come first, even though it appears in the Lens that the event is what caused you to feel the Emotions.

The story of your triggering event is a description of a condition that exists somewhere within your Domain that needs your attention. It is a metaphor for the Abscess that has become attached to a definition that is somehow preventing Emotion from fully flowing through the body. This, in turn, is preventing Intuition from fully being heard and, more importantly, acted upon.

In our model, no one "makes you mad." The triggering emotion connects you to the Abscess and provides clues to the real root of your conflict within you. As you make ready to engage your imagination and revolve your reference frame 180 degrees back to yourself, it is best to adopt the values of passion, creativity, nonjudgment, and humor.

This is the first step in your Action Plan.

Detachment: Letting go of the idea of anything outside you as the source of your thoughts or feelings and revolving your perception 180 degrees. Judgments and trapped Emotion are projected into the Lens, creating the triggering event in something or with someone that appears to be separate from you. As you detach from the idea that someone else is the cause of your Emotion and revolve your focus back to yourself, you gain the power to take control of your Domain. By redefining your world as originating within, you place dominion back in your free will. You decide how to care for your feelings, take control of your thoughts, and listen for the impelling of Compassion from within, guiding you to the next step that is good for the entire operating system, not just the ego. You become the change when you take the action to fulfill Intuition's calling.

Being detached does not mean not caring. Quite the contrary, when you revolve your story 180 degrees, detachment means to cease blaming others for your feelings and to take 100% responsibility for taking care of yourself. By taking control of your Domain, you exercise self-love, which then perfectly reflects outside of you in the form of love and the power to influence for the greatest good.

Since the physical world begins within you, by analyzing the details of the triggering event, you can create the road map that revolves 180 degrees back to your own Domain. From there, you learn where *you* are perpetrating the imbalance and take action to make new choices that honor the good of all involved—all participants in your drama, as well as all three of your Human Dimensions.

Honor and Articulate the Conflict

To experience the separation of the three Human Dimensions, assume the perspective of the Honored 4th and articulate your conflict. Write it out in detail:

1. What stories is your Intellect telling you about this conflict? What are the details? Who/what are the triggers? What have they done? How did it make you feel? If it's a situation and not a person, what predicament are you in? What judgments are you making about yourself and others involved? Who is judging you, and what are they saying? We want to see exactly how your ego is defining the situation.

Act II: Conflict REVOLUTION

Remember our values: be *passionate* and *creative* in writing about your conflict. Let yourself feel the feelings around it as you honestly and openly describe it. There are *no judgments* here, even of your judgments. In fact, we want to see your judgments so we can change them. This is where *humor* can help dispel the intensity of such deep self-scrutiny.

Keep these stories in your head region. Experience them as a part of Intellect, constantly defining and redefining the world/your life/this conflict. Separate them from Emotion, which is isolated in the solar plexus.

When you are done writing out your conflict, give your conflict a name. Part of naming the conflict is to inject some humor into what is normally a serious and confusing situation. "I am a Liar," "The Noisy Neighbors," and "The Drunken Marketing Manager" are some of my favorite names for my conflicts.

I have chosen to use my conflict "Bennie and the Slavs" as an example of how to write out your conflict and then work your way through it. The actual conflict at the time was physical: I had a distressed right shoulder, a painful lower back, and chronic insomnia. I knew, however, that no matter what conflict I chose, I would get to the root of them all. So I picked one to Rev that had more concrete details and was perhaps less vague than one about my body.

Bennie and the Slavs

At the time of this conflict, I was working as an assistant for the property manager of several vacation rentals. I was in charge of managing a three-story, six-bedroom rental property. This meant, among other things, working with the cleaning crew to make sure it was cleaned in time for the next check-in.

At the time I was suffering from chronic shoulder and lower back pain. I had tried everything from acupuncture to yoga, chiropractic to painkillers. Nothing seemed to ease the pain. Because I could not sleep on either my back or my right side, I had become an insomniac, which greatly contributed to my imbalanced emotional state. Because of that, I told my boss clearly that I was physically incapable of cleaning the rental units, so we would have to hire someone to do it.

My boss had a relationship with another rental property owner named Bennie who was utilizing a program to get cleaning help from overseas. Every year, he hired college kids from several European countries who would come and clean. In exchange for room and board, they receive a modest pay and the chance to experience American life. Many of these kids have relocated to become permanent members of our community.

My boss and Bennie made a deal that we would get to use Bennie's maids to clean our rental unit. The trouble was, Bennie was fairly underhanded in the way he treated them. They were not supposed to work more than 40 hours, but Bennie had them on the schedule for sometimes 60 hours a week. He wasn't supposed to rent them out to others or profit off them, but that's just what he was doing with us. I objected to this arrangement, but my boss overrode me and moved forward, paying Bennie in advance.

From day one, it was a nightmare. Sometimes he did not feed them or pay them as per his contract. Some weeks the only food he would provide was pizza. One day I picked them up and dropped them off at the rental house. When I returned three hours later, they were still sitting on the porch, smoking cigarettes, the house untouched. When I asked why, they told me in their broken English that they had already worked their 40 hours and had not been paid in weeks. I felt badly for them; they were prisoners with no way to escape the system until their contract was up.

When the house was not clean, who do you suppose had to take care of it? Yes, that's right, me. Me with my bad back and shoulder, even though I had told my boss that I couldn't clean. Yet time after time, because of this agreement he had with Bennie, I was left to clean up the mess he was making.

I had no say in the matter other than quitting. My boss would not talk to Bennie about it, nor would he clean when the cleaners fell through. Because of it, my arm and back were getting worse. I was not sleeping and about to go out of my mind from the physical and emotional stress. On top of this was my husband's fear of lost revenue and his pressure for me to keep my job at all costs. I grew angrier and angrier every day, knowing what was going to happen and who was going to have to clean up the mess. The pressure at home was palpable as well. I began to hate not only Bennie, but my boss, my husband, and my whole life.

This is the kind of narrative you should write about the details of the conflict. Be as real as you can. Tell the truth about how you feel about it, what you've done about it, and where it stands in present moment. This process of honoring and articulating the conflict is the first step.

The Sound Bytes

The complex and often confusing Intellect, with its ever-present imagination, has a tendency to over-analyze, judge, project, blow things out of proportion, and obsess, especially when triggered. Intellect and ego kick in with long, convoluted stories that may or may not be accurate, fueled by Emotion. This constant, obsessive jabbering of the imagination is a distraction, making anxiety abscess and preventing you from sensing Intuition. Now ego is in control, as you become trapped in your head and the entire system gets thrown out of balance.

In order to reduce the fluctuations of the Intellect and redirect thinking, we use *sound bytes*—small statements that simplify the story to a handful of easy-to-remember words and phrases. Technically, a *byte* is digital information that represents the smallest unit of memory possible. By creating a short phrase, we can easily remember our intention in the midst of all the Intellectual chaos and diversion. And not only does the activity of creating sound bytes redirect the Intellect into doing something creative, but the sound bytes then become clues in the Road Map that get revolved back to your Domain.

Time and effort will be needed on your part to figure out these sound bytes. You are working your way through the sometimes deafening and obsessive ego attached to the deeply embedded Abscess of trapped emotion. This is no easy task. Get creative until you find the perfect, simple, small statements that resonate with you and articulate the truth of your perception.

Learning to identify the three Human Dimensions separately using the details of your own life revolutionizes your operating system by cultivating self-awareness. Have fun honoring and articulating your sound bytes!

Intellectual sound bytes: From the position of Honored 4th, sift through the details of the story of your conflict. Find the obsessive, repetitive thoughts with the intention of parboiling down the themes of why you think you are conflicted about this issue. When you find the major themes, condense them into two or three simple sound bytes.

Don't be discouraged if it takes some time to wade through all the voices of culture. In the "Drunken Market Manager" conflict, it took me two weeks to even begin to try to honor and articulate the sound bytes. I was so mad over having to do the job of a drunken co-worker who was off-loading her projects onto me. Every night for two weeks on my way home from work, stuck in traffic, I would try to focus on winnowing down the obsessive, angry thoughts I was projecting onto her. It was like rounding up wild horses. I finally came up with, "She makes me do her job and then she criticizes me for it."

A similar struggle ensued with "Bennie and the Slavs." Every day I was angry and obsessed with the injustices all around me, from Bennie to my boss to my husband. It took a good two weeks from making the decision to do a Revolution before I boiled down my thoughts to this:

Act II: Conflict REVOLUTION

He lies.
They have no respect for my physical condition.
He doesn't listen to me.

Emotional sound bytes: Close your eyes, focus inward, and feel and breathe. Take a long, deep breath while thinking about the conflict. What feelings show up in your solar plexus as you breathe and think?

Look on the list of Emotions on pages 108-109. Identify just the pure, raw Emotion: Fear. Anger. Worthlessness. Anxiety. These should be one-word answers and should be experienced in your solar plexus. Pay no attention to the stories in your head of why you *think* you feel them—that is Intellect. Feel them in your abdomen.

My Emotional sound bytes were *anger, rage, helplessness, frustration,* and *worthlessness.*

Intuitive sound byte: What step might your Intuition be impelling you to take that could have a regenerative influence on the conflict? Feel this impelling bubbling up from your heart region. It might be an actual voice; it could be "goose bumps" that lead to hearing an impelling thought. Some have described it as hearing a "click" followed by clearly seeing and taking the next step.

You may struggle with this for many reasons. You are probably used to seeking authority from others—doctors, teachers, religious leaders, experts, et al. You might have little confidence or training in trusting your gut. Other times you try to take giant leaps instead of baby steps. You are looking for a small, declarative statement impelling you to take the next best baby step, one that will be good for the whole situation, including you.

Intuition could very well be telling you, "Rest," "Do nothing now," or "Feel and breathe." Your job is not to argue, but to honor and articulate the impelling of Intuition.

Emotions

Absorbed
Adventurous
Affectionate
Afraid
Aggravated
Agitated
Alarmed
Alert
Alive
Aloof
Amazed
Amused
Angry
Anguished
Animated
Annoyed
Anxious
Apathetic
Appreciative
Apprehensive
Ardent
Aroused
Ashamed
Astonished
Beat
Bewildered
Bitter
Blah
Blissful
Blue
Bored
Breathless
Brokenhearted
Buoyant
Calm
Carefree

Cheerful
Cold
Comfortable
Complacent
Composed
Concerned
Confident
Confused
Contented
Cool
Cross
Curious
Dazzled
Dejected
Delighted
Depressed
Despairing
Despondent
Detached
Disappointed
Discouraged
Disgusted
Disheartened
Dismayed
Displeased
Disquieted
Disturbed
Downhearted
Dull
Eager
Ecstatic
Edgy
Elated
Embarrassed
Embittered
Enchanted

Encouraged
Energetic
Engrossed
Enlivened
Enthusiastic
Exasperated
Excited
Exhausted
Exhilarated
Expansive
Expectant
Fascinated
Fatigued
Fearful
Forlorn
Free
Friendly
Frustrated
Fulfilled
Furious
Glad
Gleeful
Gloomy
Glorious
Glowing
Good-humored
Grateful
Gratified
Guilty
Happy
Harried
Heavy
Helpful
Helpless
Hopeful
Horrible

Act II: Conflict REVOLUTION

Horrified
Hostile
Hot
Humdrum
Hurt
Impatient
Indifferent
Inquisitive
Inspired
Intense
Interested
Intrigued
Involved
Irate
Irked
Irritated
Jealous
Jittery
Joyful
Jubilant
Lazy
Leery
Lethargic
Lonely
Loving
Mad
Mean
Mellow
Merry
Mirthful
Miserable
Morose
Mournful
Moved
Nervous
Numb

Optimistic
Overjoyed
Overwhelmed
Panicky
Peaceful
Perky
Perplexed
Pessimistic
Pleasant
Pleased
Proud
Puzzled
Quiet
Radiant
Rancorous
Rapturous
Refreshed
Relaxed
Relieved
Reluctant
Repelled
Resentful
Restless
Sad
Satisfied
Scared
Secure
Sensitive
Serene
Shakey
Shocked
Sleepy
Sorrowful
Spellbound
Splendid
Stimulated

Surprised
Suspicious
Tender
Tepid
Terrified
Thankful
Thrilled
Tired
Touched
Tranquil
Troubled
Trusting
Uncomfortable
Unconcerned
Uneasy
Unglued
Unhappy
Unnerved
Unsteady
Upbeat
Upset
Uptight
Vexed
Warm
Weary
Wide awake
Withdrawn
Woeful
Wonderful
Worried
Wretched
Zestful

Intuition won't analyze, judge, explain, or complain. It won't tell you what to do next week, next month, next year. If you find yourself explaining "why," you are still in Intellect. Intuition will impel an action about the present moment. It is not a fortune teller; it is meant to be your guide for whatever is happening right in front of you: the next most advantageous step.

The truth is, your Intellect is barely capable of understanding all the moving parts of the gynormous operating system with which you are dealing. Intellect cannot perceive what the Observer is communicating to the Source or how many events had to take place for you to end up on that street corner where you ran into just the person you wanted and needed to see. Intellect is only a small cog in the massive gears of creation, but imagination can make it appear bigger and more powerful that it truly is.

Take a moment to quiet yourself. Use your imagination to consider what a trusted friend or spiritual leader might tell you to do in your conflict. Ask the question, "What would be the next most advantageous baby step for the good of the whole, including what is best for me?"

Take a big breath, feel your feelings, and close your eyes. Get out of your head by focusing on the heart region of your body. Listen for the small, declarative statement telling you the next most advantageous step for the good of the whole situation.

When in doubt, hold the answer up against the definition. It must contain all four of these conditions to be true Intuition:

- ☯ Is the answer a small statement?
- ☯ Does the statement impel you to take an action?
- ☯ Is that action a baby step having to do with the present moment situation?
- ☯ Does that impelling step impact the whole situation for the greater good?

In a Con Rev workshop I conducted in Ohio, a participant was struggling with a conflict about her boyfriend. When I asked what her Intuitive sound byte was, she looked up sheepishly and said, "Kill him?" I calmly held it up against the model: Yes, it was a small statement. It also impelled her to take action. But it wasn't a baby step and it did not impel action that would impact the whole situation for the greater good. Even though she could imagine her life easier if he were dead, for her to kill him would wreck havoc on more lives than she could count, starting with her own. Because it did not fulfill all four criteria, we could not consider it Intuition.

After working the process long enough around Bennie and the Slavs, feeling and breathing my anger, rage, helplessness, frustration, and worthlessness, I could hear, *"Keep working the process."* Later on, farther along on the process I heard, *"Make yourself heard."*

Now that we've gotten our sound bytes articulated, we are ready to start building our personalized Road Map, Revolution, and Action Plan.

Building the Road Map

To create your Road Map, plug the information you have collected into a matrix, as illustrated on page 112. Use the name of your conflict and identify the sound bytes of your Intellect, Intuition, and Emotion. Organize them so that Intellect is at the top symbolizing your head, Intuition is in the middle representing your heart, and Emotion is at the bottom coinciding with your solar plexus.

The Revolution

Using the same creativity you called upon to create the sound bytes, begin rewriting those Intellectual statements to revolve 180 degrees. If your sound byte is about someone else, make it about you. "He doesn't respect me" turns to "I don't respect myself" or "I don't respect him."

My sound byte "They have no respect for my physical condition" became "I have no respect for my physical condition." "He lies" became "I lie." "He doesn't listen to me" became "I don't listen to myself."

Con Rev Road Map

Bennie and the Slavs	
Intellect	He lies. They have no respect for my physical condition. He doesn't listen to me.
Intuition	Make yourself heard. Keep working the process.
Emotion	Anger, rage, helplessness, frustration, worthlessness.

Create a Road Map with your sound bytes.

Act II: Conflict REVOLUTION

Another form of revolving your sound byte would be by redefining a judgment into its opposite: "I am doing it all wrong" becomes "I am really doing this perfectly." Or "I am doing it perfectly" becomes "Maybe I am making a mistake somewhere." Use your imagination and find what resonates.

Create a second map like the one on page 114 and call it "The Revolution." Change your Intellectual sound bytes to reflect the newly revolved definitions. These sound bytes are what you will be watching for in yourself.

Read your Revolution as if you were reading it for the first time. By revolving your perspective back to yourself like this, you are putting control back where it belongs. The Revolution allows you to do a deep self-scrutiny of your own thoughts, feelings, and senses.

Notice the triggering emotion when you consider that *you* could be the one who is not respecting yourself. Notice your thoughts when you realize that *you* might be the liar. Your ego's first and most natural inclination is to deny it: "No, it's not me, it's him!" or "I am not like that! I don't lie!" These denials and projections are part of the tricky ego trying to keep control, maneuvering so that you won't have to feel those feelings listed on your Road Map. If you project them onto someone else, they can appear to remain out of your control. This is how you have probably operated your entire life. No one taught you to do it differently.

In truth, you have every right and responsibility to feel all of your feelings, and only you can take control of the wild horses of your Intellect. This is the imbalance Conflict REVOLUTION® is addressing.

Rest assured, this is not about letting people off the hook for their responsibilities. I was willing to revolve my perspective of Bennie and my boss back to me even though they clearly *were* acting for the good of the few at the expense of the many. But going back to the ground rules, *my* Domain is my responsibility: I am not in charge of their Domain. My Domain is *my* responsibility: This was *my* rage, not theirs. Only I could feel it. These were *my* thoughts and

The Revolution

\	Bennie and the Slavs
Intellect	I lie. I have no respect for my physical condition. I don't listen to myself.
Intuition	Make yourself heard. Keep working the process.
Emotion	Anger, rage, helplessness, frustration, worthlessness.

Revolve your Intellectual sound bytes 180 degrees.

Act II: Conflict REVOLUTION

judgments. Only I could control and transform them. This was about me looking at myself and first and foremost being 100% responsible for *my* Domain.

Aligning all of this to Compassion requires the use of two different skill sets, each one equally important, but serving two entirely different purposes. Instead of spending your time obsessing about the characters in your drama and worrying over its details, your job is to learn these skills and inspire yourself not only to use them, but know which one to employ under which conditions.

REVOLUTION Skill Set #1: Intellect

Revolving of the Intellectual sound bytes is a journey of self-discovery. Because your defensive ego is in the habit of projection, the Revolution can start a fight in your head. As you begin to hear the awakening superconscious speaking the truth and ego feels itself losing control, brace yourself for some internal backlash.

My sound byte, "He lies," became "I lie." My ego's response to hearing this was, "I am aghast! I am not a liar! I am one of the most honest people you will ever meet! Just ask anyone!" I was outraged at the suggestion I was the one who was lying!

Aligning the Intellect to Compassion is not just about entertaining the idea that you yourself are the liar. That is the first step. You then must watch yourself throughout the day from the reference frame of the Honored 4th in order to catch yourself in the act of lying. The reason you create sound bytes in the first place is so you can easily remember what you are looking for. Imagination can run wild and create a cacophony of stories and distractions. The sound bytes cut through all that and help you focus on what you are intending to witness in yourself.

The sound byte about lying is one most people can relate to. Your ego has a habit of continually lying to yourself with a steady stream of self-talk that diminishes your power and the miraculous truth of creation happening right before your eyes. (See page 98, "Degenerative

Voices of Culture.") The lies come in the form of subjective descriptions about yourself or others, where you are either victim or perpetrator: you can't do the job well enough or you always screw up—it's your fault; or "they" always do this to you and "they" are the enemy—it's "their" fault, etc. On and on come the lies that project and keep you powerless. As you focus on these stories looking for the lies, you find repetitive themes running through them. Identify the lies and you can change the definitions to reflect the truth.

The truth is, humans have to lie in order to even exist. The truth is, we are all one being, but the human experience requires separation. Part of the mathematics in the Source that creates the ability to perceive the separation is programmed to negate the truth of unity. You simply must be separate in order to be alive as human beings. So at the crux of all creation lies a paradox of quasi-truth. It is not that separation is a lie, it's only part of the operating system.

However, I had to specifically watch for where I was lying to myself, as well as where I was disrespecting my physical condition and not listening to myself. In order to change the system, I had to catch myself in the act of perpetuating the very things that I was so outraged about in my boss and Bennie.

Understanding and exercising this skill will revolutionize the way you relate to the world. Because you are committed to observing yourself, you can now focus more in present moment and less in the tape loop obsessively running inside your head projecting onto the actors in your drama. You become more like the director of the play than the actor on the stage. Just the act of observing your thoughts and feelings and listening for Intuition gets you out of your ego and using the whole system. Intending to find those decisions that you yourself are making that are perpetuating the specific behaviors that you judge as unacceptable in someone else brings you to the doorstep of true transformation.

In "Bennie and The Slavs," I spent two solid weeks struggling to keep that intention. I argued with myself in rage about how much this was *not* about me. "I would never disrespect my body! Look at me! I have spent a small fortune on chiropractors, acupuncturists, yoga classes, pain pills! I am doing everything I can to heal my body! It's not me, it's my boss and Bennie!" I could not tame those wild horses of my thoughts as I openly and willfully broke all the ground rules of Conflict REVOLUTION®. I would get overwhelmed with anger and my mind would repeat the injustices that my boss and his associate were raining upon me. No matter how hard I tried, I could not focus on me and how I might be disrespecting my own body. Whenever I did, it just made me madder and generated more projection onto those two men.

Finally I called my friend Lily and complained that I was failing miserably at our own process! She reminded me that as long as I was still working the process with the intention to find those decisions, I was not failing, I was "in process." But having those sound bytes made it easier in those moments of internal conflict to remember what I was looking for. Lily helped me remember that I was looking for any time that I was making a decision that was not in my body's best interest.

REVOLUTION Skill Set #2: Emotion

To align to Compassion, Emotion becomes an energy that is processed through the body using breath, not an intellectual analysis of what might have "caused" the triggering event. Yes, those details are needed to create Intellectual sound bytes. But analyzing the details doesn't help to release deep feelings that have been trapped for years in the Abscess. Intellectual analyzing, in fact, can prevent the release of deep feelings by keeping Emotion attached to and fueling the story of the triggering event. Right now, you really need to get out of your head and into your body to feel those feelings, not think about them. By focusing instead on thoughts that assist the breath to move those feelings through your body, you create an intentional change.

Emotion is your passion, the creative rocket fuel that funds your visions. You *want* to feel the full range of feelings—even the alleged "bad" ones of anger, fear, depression, etc. Remember, what would the color wheel be without red? But instead of allowing your feelings to automatically fuel thoughts of denial and projection, you use your breath to move all of your feelings through your body and free them at last from their prison in the Abscess. By using superconscious thoughts to intentionally help your feelings flow through your body without judgment or attachment, you change the operating system, right before your eyes.

Feel and Breathe. Intentionally stop and take five deep, calming breaths. Feel Emotion in your solar plexus flowing up from the Source, through your body, flowing out into the heavens through the top of your head. Feel Emotion as electricity that is becoming one with the Earth's electromagnetic field and being drawn up to the North Pole and back down into the center of the planet, back to the Source, only to begin again.

Compassionate Intellectual sound bytes. As you use your breath to move Emotion, attach it to a nonjudgmental thought that will help keep Emotion moving. These are the Compassionate Intellectual sound bytes to use to get your ego out of the way and allow Emotion to flow.

Experiencing Emotion as electricity also removes the stories of Intellect. You can also create a sound byte that becomes your mantra. Repeat when your deep and difficult feelings won't detach from a degenerative definition. "I don't have to know why I am feeling (fill_in_the_blank), I just have to feel and breathe. Breathe." Then take a deep breath. Creating an Intellectual statement that will help keep your feelings flowing allows you to process it *physically* instead of *intellectually*. In the process, you release more of the Abscess. That brings with it more Intuition, continuing to impel you to feel and breathe, as well as to keep reflecting on your Intellectual sound byte and taking baby steps for the good of the entire situation.

Notes

Act III

The Action Plan

Action Plan

Bennie and the Slavs	
Intellect	I will listen to my body and my Intuition; I will pay attention to where I might be lying to myself or others; or being inauthentic. I will find out how I am not respecting my current physical condition and take action to change that.
Intuition	Find a way to make myself heard by my boss, as well as listen to myself. Keep working the process.
Emotion	I will breathe the anger, rage, helplessness, frustration, and worthlessness through my body as a way to unblock the Abscess.

Act III: The Plan

The Action Plan

The goal of the Action Plan is to create a tool to help you catch yourself perpetrating the action of your Intellectual sound byte. You're watching for the very moment when *you* are making the decision for the good of the few at the expense of the many. Then, applying your new skill sets, you will be intuitively guided as to what is the next most advantageous step to take for the good of the entire situation. At that point, you—and only you—will decide if you will take that step.

If you choose to take one baby step at a time, your Action Plan will lead you to manifest an outcome that cannot be attained by merely addressing the details of the drama of the conflict.

In "Bennie and the Slavs," the details of my drama were that I had a bad back and shoulder, a thoughtless boss and an incompetent business associate, and overworked and underpaid maids. At every turn I was in pain and raging against the next uncompleted cleaning job. No matter how many times I told my boss I could not clean, each time the maids didn't show up, I ended up not just having to clean, but having to clean under pressure. I was left holding the vacuum with a very short turn-around time to clean a three-story six-bedroom four-bath house before the next guests arrived. My boss refused to hire another cleaner, as he had already paid Bennie and was trying to get him to honor his end of the deal, which Bennie clearly was not doing.

Armed with my revolved Intellectual sound bytes and recognizing the anger, rage, helplessness, frustration, and worthlessness I needed to be feeling and breathing, I set about to create my Action Plan. Since my Revolution revealed that I was not listening to myself and was disregarding the physical condition of my body, the Action Plan became about righting those imbalances—even if I didn't yet know how I was going to do it.

I began each morning before I got out of bed. Focusing on my body, I took a deep breath and said, "Thank you for another day of life." Taking another breath, I set my intention to being a peacemaker. I committed to spending the day watching myself. Every time I found a conflict, I pledged to do my best to use Con Rev to address it.

I reminded myself of my Emotional sound bytes and committed to taking time throughout the day to breathing in those feelings as they arose. I decided on "I don't have to understand why I am mad, I just have to feel and breathe" as my Compassionate Intellectual sound byte. I would use it if and when I got triggered throughout the day. I set my intention to witness where I was lying, not respecting my physical condition, and not listening to myself.

For several days, all I did was watch myself walking around in pain, sleep-deprived, angry, and railing on Bennie and my boss in my head. But at least I was witnessing. I was also able to start breathing and releasing some of the Abscess. It didn't stop the wild horses of my Intellect from having thoughts of Bennie and my boss, but it did create a tiny space between them and the fuel of the rage. This weakened the system enough for Intuition to be heard: "Keep working the process."

I kept breathing and watching. At the same time, like a mantra, I kept asking to be shown: "Show me where I am lying. Where am I disrespecting my physical condition? Where am I not listening to myself?"

My first AHA moment came while witnessing the argument in my head about how *"I do not lie."* Clearly, that was a lie. Dang.

Catching yourself in the act of your sound byte leaves you speechless. Right before your eyes, you see yourself doing the very thing you insist that you aren't doing and are complaining about in someone else.

This revolutionary moment is called *Hangin' in Dang*. All internal arguments cease as you witness the irrefutable truth. "Dang ..." is about as much as can be heard rattling around in your stupefied ego.

Act III: The Plan

* * * * *

"I Am a Liar"

In 2002, I had a business partner, Jordan, who would lie about even the smallest things. She was bright, creative, and daring, and I often loved working with her. But I never understood why she had to lie about stuff she didn't have to lie about—her hometown, her father's occupation, what time her daughter was due home from school, etc. When I caught her in a lie, she would rationalize, weasel, or finesse her way out of it. No matter how many times I caught her, her lying never stopped, and it never stopped me from getting involved with her.

One day she suggested we plan a long weekend to work on our latest project. She even offered to pay for a room. We decided on a B&B my friend had just opened near her.

Jordan told me she had booked a room, but the owner called to say her credit card was declined and she wasn't returning calls. When I told this to Jordan, she apologized and said she would give them a different card number. An hour later, the owner called to say that Jordan had emailed a different card number but had left off a digit and was still not responding.

Finally, I called Jordan and angrily confronted her—what was going on? She again apologized and said she would definitely take care of it. Not to worry, it was a done deal; she would use her husband's card.

Sure enough, my friend called with the bad news: that card did not go through either. I thought I was going to burst a blood vessel. Instead, I took my rage and the sound byte "she is a liar" to the ski hill.

I must have skied for hours, listening to the arguments in my head about Jordan and feeling and breathing rage, sadness, and indignation. *I am not a liar*, I insisted, not like

Jordan. *She* was clearly the one lying, not me! More breathing and reminding myself of the ground rules: 1. *My* Domain is my responsibility. 2. My Domain is *my* responsibility. 3. See 1 and 2.

I clearly remember hanging there in dang as if it were yesterday. I was skiing down the big hill at French Regional Park in Minneapolis on a cold January evening. Who is a liar? I am a liar! Dang.

Suddenly I saw that I was lying to myself about Jordan in order to continue to form business associations with her. I *knew* she lied. My friend at the B&B confirmed it. But I had known long before that. I had witnessed her lies for years. She would lie right to my face without missing a beat. And yet, when it came time for another project, I would conveniently develop amnesia about her lying and the resulting havoc, as if it had never happened. I would then make a commitment to work with her again, remembering only the fun and giddy creative high we could get ourselves into. When she'd inevitably lie again, I would become outraged, as if I had not been through this many times before. What part of me was lying to myself about her? Did I really want to be in business with someone I knew for a fact would lie to my face?

Fool me once, shame on you. Fool me several more times, it's time for a Revolution.

As I caught myself lying to myself about her, the next obvious question was, where am I lying to other people? (More silence.) And what else am I lying to myself about? (Even more silence.)

Suffice it to say, "I Am a Liar" got a lot of mileage. To this day, it is a constant reminder of what the tricky ego is capable of. Today I admit, openly and without shame, I *am* a liar and I know it. I tell myself I am too old to accomplish something (which I am not), that I don't feel a particular feeling (which I do), or that someone else is the source of my pain (which they aren't). My imagination can construct countless scenarios that can seem very real. Discerning the truth can be daunting when faced with the complexity of being human. Conflict REVOLUTION® provides a process to take responsibility for my lies and change my system to align to the truth.

Once I stopped lying to myself about Jordan, I started making different decisions about my level of involvement with her. Soon she disappeared from my life completely. The last I heard about her, I got a call from a woman in Florida who had found my name associated with Jordan's on the Internet somewhere. She called to say that she had paid Jordan to do some work and that she had taken her money and run. I felt bad for the woman, but I was particularly happy to no longer be lying to myself about Jordan.

*****

By the time "Bennie and the Slavs" was playing out, I already knew I was a liar. So *this* was my first AHA moment: I am a liar? Dang! Not again! Seeing myself lying again made me even madder, which forced me to breathe harder and chant my Compassionate Intellectual sound byte like a mantra. Witnessing the irrefutable proof of catching myself lying humbled me enough to hear Intuition say: "Keep watching!"

After that demonstration, what else could I do? One byte down, two to go.

I kept watching and breathing. Just because I was witnessing my thoughts did not mean I had any control over them yet. It was excruciating to listen to the rants in my head about Bennie and my boss, but I could not stop. I was making myself sick fueling that tape loop with all my anger.

One Sunday morning, even though I had called him the night before and arranged to pick up the maids, Bennie came out of his office on Monday to cheerfully greet me at my car and tell me that it was their day off. What? What about our conversation yesterday about how I was going to pick them up this morning? He was a psychopath! It was infuriating!

I peeled out of his driveway in a rage. Racing to the rental house, I screamed my head off. The insanity of Bennie's action drove me over the edge as I demanded the universe show me where I could possibly be perpetrating this!

Flying around the house like a dervish, I watched myself raging as I made beds, did dishes, scrubbed bathtubs. While I cleaned, in my most arrogant and cynical tones, mocking even, I dared the universe to show me where I was not respecting myself! That was followed by the complete inventory of everything I had done in the past year to heal my bad arm and back. Then on I'd rage.

This was followed by a laundry list of offenses Bennie and my boss had perpetrated against me, causing all this rage! Whenever I could in the midst of the chaos, I would stop and breathe and chant my mantra. I reminded myself of the ground rules: my Domain is my responsibility. I could only barely hear Intuition urging, "Keep working the process," so I did.

It wasn't until I was cleaning the windows that I had my other two AHA moments.

As if in slow motion, I watched as I picked up the paper towels with my bad arm and proceeded to wash the window. I was the one spraying the Windex. I was the one putting more and more pressure on my arm as I got angrier and angrier as I washed. And I was the one in great pain.

While it was true that Bennie had lied again about the maids being available, Bennie was not standing there with a gun to my head, forcing me to work my bad arm.

This AHA moment was a two-fer: I watched myself not listen to my body, which in turn was completely disrespecting my physical condition. Dang and double dang.

On top of that, I was lying, insisting "I do, too, respect my physical condition!" while I continued to clean as if I weren't in pain. On top of that were many arguments about why I *had* to keep cleaning: because no one else would and I didn't want to have to face angry guests.

All those internal arguments flew right out of my head as I saw the irrefutable truth. My Honored 4th proved that I was, indeed, the liar who was making decisions that were not respecting my physical condition or listening to my own body that was crying out in pain.

Act III: The Plan

The moment you catch yourself taking action for the good of the few at the expense of the many is the very moment the system changes. Courageously facing the truth, you only need to keep feeling and breathing while you listen for Intuition to reveal the next most advantageous baby step you can take, right here, right now, to move the conflict into a resolution for the good of all. It is then up to you to take it.

Putting down the Windex and paper towels, I sobbed. When I asked, "What should I do?" my Intuition immediately said, "Make yourself heard."

First and foremost, I had to listen to my body. I had to admit that I was the one who had agreed to clean. I might have been telling my boss I wouldn't clean, but that was clearly a lie. Every time it happened, I cleaned. That was the truth. This had to stop immediately, despite how the guests might respond or if anyone else would take care of it.

I decided to drive back into town and go straight to my boss. I asked him to step outside the office because I had something very important to say to him and I didn't want the conversation to get lost in all the other business at hand. He followed willingly.

Once he was sitting in front of me on the picnic bench, I literally grabbed his ears and pulled his face very close to mine.

"I don't clean. Period. I know I have been saying this for months, but I am a liar. Even though I say it, I always end up cleaning. Well, this is to serve you warning: I will not clean anymore. You will have to manage the maids with Bennie. I will not be party to that incompetence. This means that if the maids don't clean, you will have to do it, or find someone who will, but it won't be me. If the guests show up and the house is not cleaned, I will give them your cell phone number. I need you to repeat after me: 'Barb does not clean.' Say it."

Sure enough, he said it several times and guaranteed me that he not only heard me, but understood the consequences. He, too, was frustrated with Bennie for not fulfilling his end of the bargain. He agreed we would cut our losses and find someone we could rely on. A great weight lifted off my shoulders in that moment but we still had no idea who that would be. It was the height of the season and good help was impossible to find.

As soon as we got back to the office, a woman walked in and said she heard we needed a cleaner. We hired her on the spot and it was never an issue again. Betty was a hard worker and showed up every single time.

This is what Conflict REVOLUTION® allows you to manifest: an outcome for the good of the whole that you might not be able to achieve just by trying to manipulate the details of the conflict. When you move trapped emotion out of the Abscess and change your thinking, the relaxed mind and newly flowing Emotion make space for Intuition to rise up and be heard. Unplugged from Emotion, the calmer mind is more willing to take the step Intuition is impelling. This is Cn^3. When we align to Compassion in this way, miraculous things can happen.

Let's redefine "miracle." What could be defined as "a rare event caused by a supernatural being" now becomes "the byproduct of the conscious intention to act for the good of the whole that impels mysterious potentials to manifest, almost as if by magic." Betty walking into the office the very day I revolved myself is the kind of "miracle" we see when using this process.

Monumental change does not have to mean turning water into wine or raising the dead. It can mean a sudden change in a conflict that has played out the same way for years or the day you finally stand up for yourself and speak the truth after years of remaining silent.

Direct and Indirect Mirrors

Conflict REVOLUTION® is about looking in the mirror and finding where we are perpetrating the actions that we see outside ourselves. With "Bennie and the Slavs," I was clearly projecting my own actions onto Bennie and my boss. Had I not witnessed for myself, I never would have known that I, too, was acting like they were, in my own way. So Bennie and my boss were *direct mirrors* for me.

Act III: The Plan

Not all conflicts are direct mirrors. Some Revolutions will lead you to a different arena than the one in which the details of your conflict are playing out. In the case of "Drunken Marketing Manager," my conflict was about a woman with whom I worked. My sound byte, "She makes me do my work and then criticizes me for it," was unbeknownst to me playing out in a different arena than the workplace and with a different person than her. This is an *indirect mirror.*

Early in my marriage, my husband and I went through a rocky time. In a trial separation, I had moved 500 miles away and stopped paying my share of the bills and participating in the upkeep of our home. I said I left him because he didn't support me. And yet there he was, all alone, paying all the bills, putting up the storm windows, taking care of our lives, while I was off blaming him for not supporting me.

We weathered the storm and were back together in our home when the drunken marketing manager was driving me insane. When I had my AHA moment, I suddenly witnessed myself back in that time, expecting my husband to do all the work and still criticizing him for it. Dang.

After hanging there in dang for a while, I went to my husband and made amends. I had never realized the injustice of my actions. The day after we had a long, healing talk and I took responsibility for being so unfair to him back then, the drunken marketing manager was fired.

In the case of "The Noisy Neighbors," it took over a year of Revving, but the end result was that the loud, drug-dealing neighbors were finally evicted and my home became quiet and peaceful again. My sound byte, "They have no idea the impact they are having on me" revolved to, "I have no idea the impact I am having on me" which caused me to examine the times I underestimated my own power. "I have no idea the impact I am having on others," uncovered actions I had taken in the past that affected situations for the good of only me. My argumentative responses, my selfish decisions, the many times when I wanted to get back at someone—a cornucopia of AHA moments revealed themselves that, one by one, I could feel and breathe my way through, forgive myself, empty some more of the Abscess, and make an Action Plan to change. By the time the eviction arrived and the neighbors were raging downstairs, blaming me, I was overjoyed to have also dealt with many other unresolved conflicts from my past that no longer plagued me. When the "Noisy Neighbors" finally left, my home became peaceful, in more ways than one.

After Betty was hired that summer, all my conflicts certainly did not disappear. But as I continued to use Con Rev to find the times I was lying and disregarding myself, I discovered many more situations where I was ignoring my own health, well-being, and happiness. Eventually, I revolved and resolved enough conflict to reach the fundamental questions: what is my wildest dream and what is getting in the way of making it come true?

My heart really wanted to be training Conflict REVOLUTION® and promoting *Imagining Einstein*. Yet there I was with this brilliant new book sitting on the shelf, in a bad marriage, and working as a realtor. I made the decision to risk creating the life I wanted to live, not the one my ego was telling me I should.

Having no idea how I was going to do this, I separated from my husband, moved to Texas, and began promoting my Einstein work. While I muddled along one step at a time, serendipity stepped in when I was invited to appear at an event in September 2007 that was broadcast internationally to hundreds of thousands of people worldwide. I was a huge hit with the Crimson Circle and was invited all over the world to train Conflict REVOLUTION® and channel Einstein and the Party. I could not have planned it any better.

"Bennie and the Slavs" was originally inspired by my physical conflict: my bad shoulder and back, and my chronic insomnia. By September 2007, my physical pain was gone and I was well on my way to making my wildest dreams come true.

Creating Your Unified Field, One Step at a Time

1. Before you get up in the morning, say a prayer of gratitude to be blessed with another day of life. Then pay attention!

2. Set an intention to be in Honored 4th and witness your Intellect, Emotion, and Intuition as separate energies throughout your day.

3. Feel and breathe whatever Emotion is moving through you in present moment, without attaching it to a story. Remind yourself of your Emotional sound bytes and create new Compassionate Intellectual sound bytes that support processing Emotion by moving it through the body with breath.

4. Work to master the wild horses of your Intellect. Find ways to quiet the mind, such as yoga and meditation. Don't wait for a special time of day or a class to meditate; use the "unimportant" times, like when you are waiting in line at the supermarket, to quiet the mind and focus on the here and now. Check in with your solar plexus and feel and breathe. Do a quick 30-second scan from head to toe to settle down your thoughts. Step into Honored 4th and observe them with detachment.

5. Carry your Intellectual sound bytes around with you. Ask that you be shown where you are perpetrating the conditions articulated in your sound byte. Pay attention throughout the day, watch for when and where you are making or have made decisions that are creating these conditions..

6. When you have an AHA moment and catch yourself in the act of perpetuating your Intellectual sound byte, feel and breathe all triggering emotion through your body. Use passion, nonjudgment, creativity, and humor to stay in the moment as the wave of Emotion passes.

7. Make a commitment to your Intuition. When you have an AHA moment, ask Intuition what the next most advantageous step is for the good of the whole situation. Then take it. In case of doubt, use this rule of thumb: If Intuition isn't emphatically indicating a step, do nothing for the moment. It's just one step. Keeping listening and feeling and breathing.

8. Pay attention to what you are manifesting. Make a list of the small changes that are beginning to manifest in your Lens. Notice times when you are less attached and can easily get out of your head. Be aware of the moments when you make yourself feel and breathe. Embrace the difficulty and discomfort of moving anger or anxiety through your body, but notice how much better you feel when the wave passes.

Keep a journal and document the changes happening outside you. Pay attention to when a conflict "magically" resolves itself or when there is movement in an otherwise insurmountable block.

9. Celebrate the baby steps! Even though the changes might not resolve the entirety of the conflict, every step is part of the millions of baby steps you will take to do just that. Don't minimize the baby steps, celebrate them!

Tombstone
Pax Vobis Cemetery
Prague, Czech Republic
November 2014
Photo: Barbara With

Einstein in Prague

In 1911, Albert Einstein took a teaching assignment in Prague as a full university professor of theoretical physics at Charles University. By now, he had already achieved rock star status with his special theory of relativity and other successful studies in thermodynamics and molecular physics.

Prague had a vibrant and thriving Jewish community. Over half the Jews in the city spoke German, which provided Einstein with an outlet for his penchant for philosophical and literary debating. Soon he was frequenting the home of Bertha Fanta, where he encountered the likes of writers Max Brod and Franz Kafka, philosopher and Zionist activist Hugo Bergmann, and where he first met a man who would become his colleague and dear friend: Max Planck.

Bertha's salon was also a musical venue where Einstein enjoyed playing his violin with several piano players in the group. This was of great importance for him, for he found stimulation not only in music but in debate, a well-documented part of his history, wherever he was.

Prague eventually fell to the Nazis, and 95% of its Jewish population was sent to death camps. Einstein fled Europe for the United States in 1933 and eventually arrived at Princeton University. No doubt he was profoundly affected by having to watch his wonderful consort of intellectuals and musicians face the genocide that Hitler was perpetrating across Europe.

If there was a way to reach from beyond the grave, surely Einstein could not only find a way to do it, but he could bring back a formula for others to follow that would lead to what he calls "world peace, one person at a time, starting with you." He could very well have carried his passion for global peace into Afterlife.

Why not?

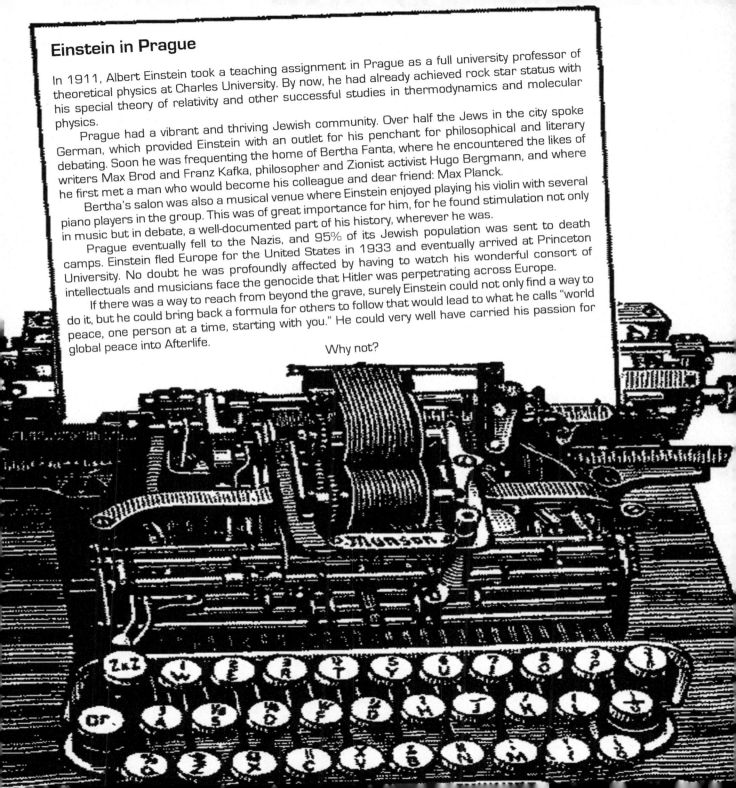

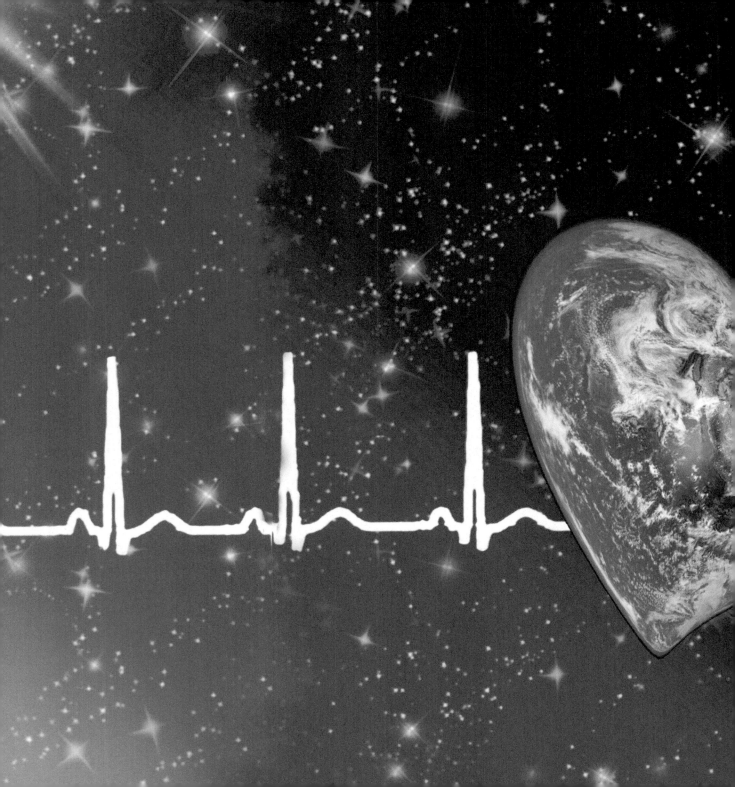

Finale

Living Aligned to Compassion: Cn^3

World Peace and Making Your Wildest Dreams Come True

When you are aligned to Compassion and achieve Cn^3, Emotion flows freely through you, bringing with it the voice of Intuition telling you the next most advantageous step for the good of the whole. From the reference frame of the Honored 4th, Intellect observes the details of what is being created in present moment in the Lens, as well as listens for the Intuitive impelling. Then, like a faithful servant, Intellect uses free will to fulfill the Intuitive desire. If Intuition says, "Rest," Intellect immediately finds a way to rest—whether it's mind, body, or spirit. Intellect does not need to know why you are being asked to rest; it only needs to do it, in complete trust that Intuition is guiding the entire system to manifest an outcome for the greatest good for all.

When the three Human Dimensions are working in concert with each other, you create inner peace. This inner peace can't help but reflect in the world around you.

Every decision you make creates your life. Imagine how many decisions you make in one day. The more you cultivate the Honored 4th, the more control you have to use free will to make decisions aligned with your Intuition. The more you feel and breathe all Emotion, the more you can align your perspective to Compassion and naturally manifest exactly what you need at any moment in the Lens of your physical world. You are also emptying out the Abscess and releasing Emotion that has been trapped, perhaps for your entire life.

As you make decisions for the good of all, the math of your actions becomes your new gravity. Traveling out into the heavens and back up through the north pole, your string begins again when it leaves the Source with this new math. This then changes the embedded descriptions in the subatomic. "If I get mad, then I get hurt" becomes, "I have every right and responsibility to feel all my feelings," because that is what is the best for you, an important part of the whole. In fact, without "you" there would be no ability to observe the physical universe. Without "you," there would be nothing by math.

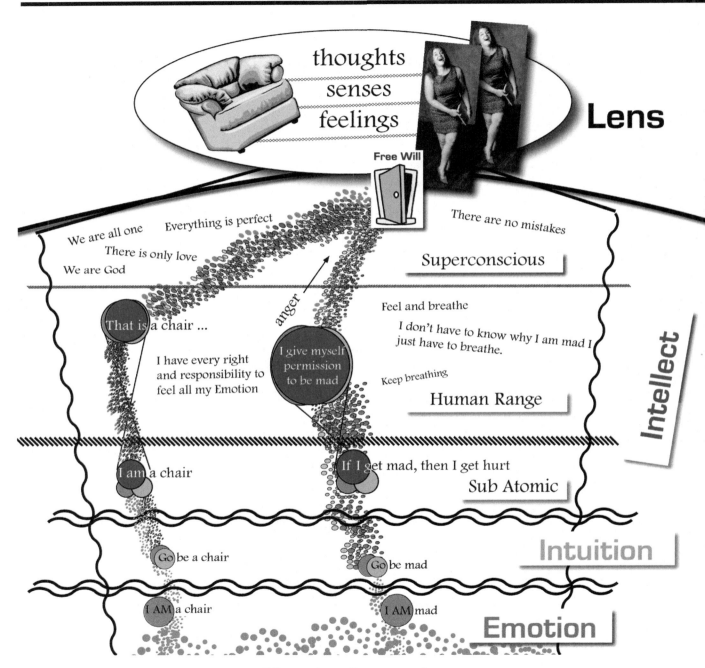

Imagine! What if every person in the entire world took responsibility for his or her own Domain? Energy that had been used in creating conflict would be freed up for creating peace and working for the good of all. People would tap into hidden talents they never knew they had. Everyone would be self-contained, self-loving, and self-monitoring; they would understand what they need and know how to fulfill those needs. Oh, how much different the world might be!

When you take full responsibility for your Domain and nurture yourself, you become a person of action instead of reaction, in charge of your own care and feeding. You become your own Source, of everything! Talk about control!

Remember, no matter what, *you always have a free will choice* as to how you will respond to what is happening in present moment. By stepping back into Honored 4th, examining the situation, listening for the next most advantageous step and then taking it, you truly become the master of your own life.

Aligning to Compassion conserves and redirects your life force and alters the direction of planet Earth towards regeneration and into disarmament. By disarming yourself, you find the path to true power, the kind that recognizes and supports all the parts. Imagine having four children, and only feeling obligated to support two of them. Surely your family would suffer, in more ways than one. Humanity's challenge is finding ways to honor that everyone is one and everything is a miraculous, unexplainable creation of the mysterious force of Compassion.

Conflict REVOLUTION® is not about sitting on a mountaintop, away from all the cares of the world with nothing to do but meditate. Sustaining intimate, intricate self-awareness on a daily basis is one of the hardest things you can do, especially when you feel angry, frustrated, or threatened. The imbalance between Intellect and Emotion can have a tight grip on you. But persistence, hard work, and a strong commitment to your own transformation also changes the mathematics of your gravity. Once you've changed on this root level, choosing Compassion becomes as knee-jerk as perceiving separation had been.

Now when Emotion rises, an entirely new set of actions kicks in, causing new reactions. Intellect knows to protect the world from its projections. It tunes into the messages of Compassion and uses them to support feeling and breathing Emotion. With nonjudgmental messages of truth and unity from the superconscious, you are reminded that, as you breathe in Emotion and embrace it, it becomes a part of your unity. Now you are a present-moment, interactive unified field. You are a living truth, accepting your power, committed to and capable of contributing to world peace like never before.

If you want to be physically healthy, you create a routine of healthy eating and exercise. If you want to be spiritual, you routinely meditate and pray. If you want to be in politics, you routinely devote time to political causes. If you want to revolve your conflicts in this way, you take time to self-scrutinize on a daily basis.

Compassion's intention is to create a physical world that operates for the good of all the parts. These are the roots of evolution and absolutely necessary for the survival of the human species. But none of the parts of the whole alone can save the whole world. Survival won't happen just by changing politics, although it's good to work for political change. The evolution of the human species is not just changing religion or science, but certainly changes will be made there as well. What will save the world will be the parts coming together to accept and embrace a new definition of the whole. When enough individual humans begin to ask how to create peace instead of how to prove they're right, then the world will shift on the quantum level.

If you can't get Intellect under control, step into Honored 4th and use your creativity to follow the clues to your own Road Map. Define what you think is the cause of your conflict, then set about to witness where you might be perpetuating these very conditions.

This kind of accountability and consciousness of your consciousness is beyond science and religion. It is beyond politics and business, beyond humanitarian efforts. It is a transformation of an individual compilation of consciousness on a molecular level, using free will to integrate the parts into the whole to create Cn^3. That will manifest heaven on Earth.

And that, dear friends, is the next frontier.

Notes

The GIFTS of Conflict

 **ratitude**
Thank the universe for bringing you another day of precious, mysterious, miraculous life. It is an opportunity to get to know yourself and become the change.

 **ntention**
Intend to find your inner conflicts and revolve and resolve them.

 **orgiveness**
To forgive is to grant pardon without harboring resentment. This is love.

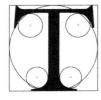 **enacity**
Have the tenacity to continue to revolve your focus. Commit to the mystery.

 **elf-Love**
The ultimate goal is to be self-responsible and self-regulating. This is the action of actively loving yourself.

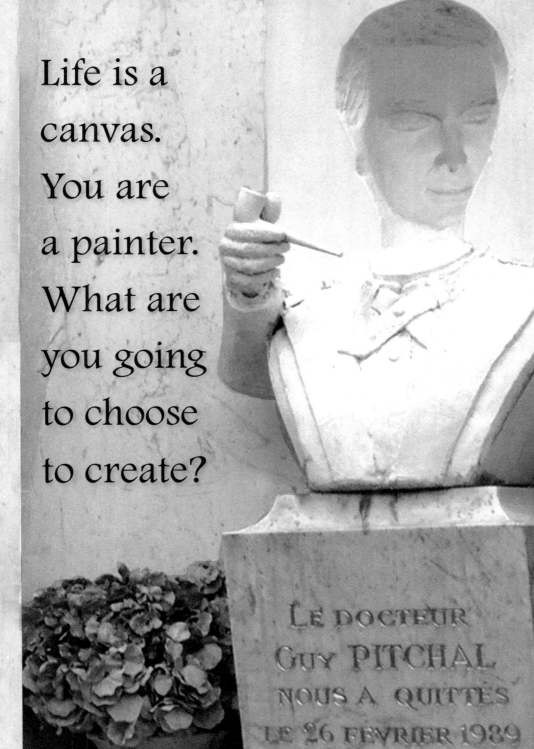

Life is a canvas. You are a painter. What are you going to choose to create?

About Barbara With

Barbara With is an international peace activist, award-winning author and composer, performer, psychic, workshop facilitator, and inspirational speaker living in northern Wisconsin. Her other books include *Imagining Einstein: Essays on M-Theory, World Peace & The Science of Compassion*, winner of the National Best Books 2007 Award for Fiction & Literature: New Age Fiction and the 2007 Indie Excellence Book Award for New Age Fiction; *Party of Twelve: The Afterlife Interviews*, winner of the 2008 Best Beach Book Award for Spirituality; *Guerrilla Publishing: How to Become a Published Author & Keep 100% of Your Profits*; *Madeline Island Artist Colony*; *UNINTIMIDATED: Wisconsin Sings Truth to Power*, runner up in the 2013 Great Midwest Book Festival: Photography; and *Diaries of a Psychic Sorority: Talking With the Angels*, coauthored with Teresa McMillian and Lily Phelps. She also has two CD of original music, *Innocent Future* and *Solitaire*. Her song, *Voices in the Wind*, co-written with Raeanne Ruth, was runner up in the 2001 John Lennon Songwriting Contest.

Together with Lily and Teresa, she researched and developed Conflict REVOLUTION®, a revolutionary new way for resolving conflict on a root level of human consciousness. Barbara is available to train and coach Conflict REVOLUTION®, conduct private and group channels, and/or perform her original music. Contact Mad Island Communications for more information.

www.barbarawith.com
partyof12.com
barbwith.com
barbarawith11@aol.com
715.209.5471